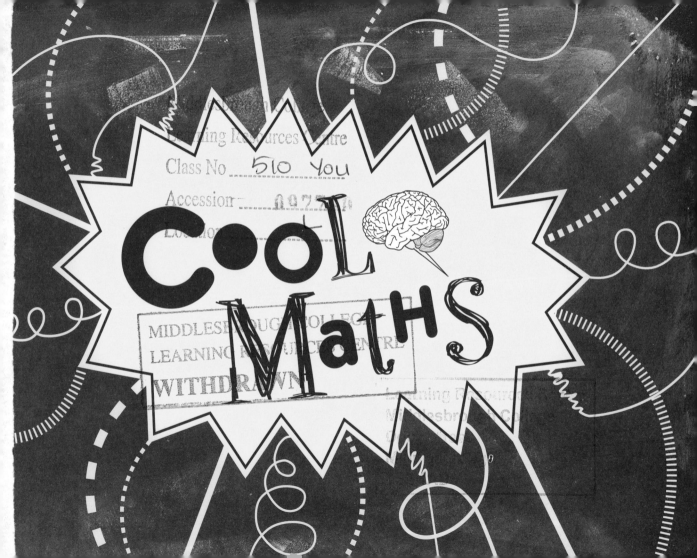

CooL MatHS

First published in the United Kingdom in 2013 by
Portico Books
10 Southcombe Street
London
W14 0RA

An imprint of Anova Books Company Ltd

ISBN 9781907554841

A CIP catalogue record for this book is available from the
British Library.

10 9 8 7 6 5 4 3 2 1

Printed and bound by 1010 Printing International Ltd, China

This book can be ordered direct from the publisher at
www.anovabooks.com

CONTENTS

Introduction

The Three Rs – Reading, Writing and Arithmetic – are the mainstay of every education system, and as such always come across as a little dull. J.K. Rowling has done her bit in recent years to make reading cooler and more fun, and has even inspired many young people to try their hand at writing, and that's great. But what about poor old arithmetic?

Well for one thing, these days no one calls it arithmetic. No one even calls it mathematics any more, it's just plain and simple maths – or if you are American, math.

Okay, maths may be plain and simple to people who are not willing to understand it, but to others it is a place where magic and wonder exists within numbers and equations, theories and formulas. It can bring the ordinary world to life in new, exciting ways and turn a humdrum situation into one of endless, impossible, improbable fun.

Yes, you did just read the words MATHS and FUN in the same paragraph, and now you've just read them in the same sentence, but how can that really be achieved? Here's how …

Maths is not just 2 + 2 = 4. Maths can help you predict the outcome of seemingly random events and the improbable. It can also allow you to find out how high Big Ben is, for example, without even measuring it. Maths can let you do the impossible and it can confound all your expectations.

Maths is everywhere. It is in everything we see, feel, know and do. It is the fundamental understanding of the importance of all types of maths, from geometry to trigonometry, calculus to probability, that has enabled man to walk on the Moon, send robots to Mars, allows technology to work back here on Earth and, most importantly, maths – and the process of computing and analyzing data – is how your brain gets you from A to B.

Leave your doubts at the door and jump into the world of *Cool Maths*.

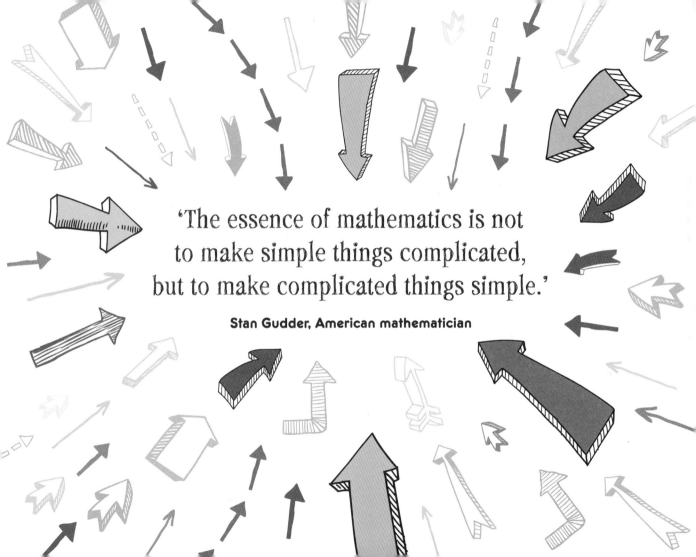

'The essence of mathematics is not to make simple things complicated, but to make complicated things simple.'

Stan Gudder, American mathematician

Great Moments in Maths

About 30000BC
Palaeolithic peoples in central Europe and France record numbers on bones.

About 450BC
Greeks begin to use written numerals.

263 By using a regular polygon with 192 sides Liu Hui calculates the value of π as 3.14159, correct to five decimal places.

About 3000BC The abacus is developed in the Middle East and around the Mediterranean.

About 300BC Euclid gives a systematic development of geometry in his *Elements*.

594 Decimal notation, the system on which our current notation is based, is used for numbers in India.

1950–1750BC
The Babylonians (from part of present-day Iraq) know linear and quadratic equations, multiplication tables, square and cube roots.

About 240BC Archimedes produces his inventions, including the Archimedes screw, and his writings on mathematics.

About 980 French scholar Gerbert of Aurillac (later Pope Sylvester II) reintroduces the abacus into Europe. Uses Indian/Arabic numerals without a zero.

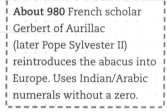

575BC Greek mathematician Thales brings Babylonian mathematical knowledge, including geometry, to Greece.

200BC Eratosthenes develops his sieve to isolate prime numbers.

About 1AD Chinese mathematician Liu Hsin uses decimal fractions.

1150 Arabic numerals are introduced into Europe with Italian mathematician Gherard of Cremona's translation of Ptolemy's *Almagest*.

500BC Pythagoras and his school, the Pythagoreans, study irrational numbers, the Golden Ratio, properties of triangles and Pythagorean theorem.

1202 Italian mathematician Fibonacci writes *Liber abaci* and calculates the Fibonacci sequence.

1494 Italian mathematician Luca Pacioli publishes *Summa de arithmetica, geometria, proportioni et proportionalita*, a summary of all the mathematics known at the time.

1514 Dutch mathematician Giel Vander Hoecke uses the '+' and '-' signs.

1557 Welsh doctor and mathematician Robert Recorde publishes *The Whetstone of Witte* that introduces '=' (the equals sign) into mathematics.

1591 Frenchman François Viète uses letters as symbols for known and unknown quantities. Descartes later uses the letters 'x' and 'y' for unknowns.

1615 German mathematician Johannes Kepler publishes work that shows early use of calculus.

1626 French mathematician Albert Girard publishes a work on trigonometry containing the first use of the abbreviations sin, cos and tan.

1665 English mathematician Isaac Newton discovers binomial theorem and begins work on differential calculus.

1687 Newton publishes *The Principia* or *Philosophiae naturalis principia mathematica* (*The Mathematical Principles of Natural Philosophy*).

1794 Frenchman Adrien-Marie Legendre publishes *Eléments de Géométrie*, an account of geometry that is a leading text for 100 years.

1799 Metric system introduced in France.

1823 Englishman Charles Babbage starts to build his 'difference engine', capable of calculating logarithms and trigonometric functions.

1879 English mathematician Alfred Bray Kempe publishes his false proof of the Four Colour Theorem.

1976 Americans Kenneth Appel and Wolfgang Haken show that Kempe's Four Colour conjecture is true.

1994 English mathematician Andrew John Wiles proves Fermat's Last Theorem.

2003 Russian Grigori Perelman proves the Poincaré conjecture relating to 3-D spaces, first proposed in 1904, by Henri Poincaré.

Multiplication Made Easy

There's always someone who knows the tricks of the trade. You know, like the man who knows a good way to change a spark plug, or the guy who can reboot his washing machine when it breaks down. Multiplication is no different and here are a few little tricks to make life easier.

Multiplying by 9

Multiplying by 10 is simple: you just add a 0 at the end. If only multiplying by 9 was so easy. Well it can be. Here is a super tip for multiplying any number from 1 to 10 by 9.

Let's Work It Out!

Hold your hands in front of your face with your palms facing away and your fingers outstretched.

Starting from your left-hand side, whatever number you want to multiply by 9, bend that finger down. So if you want 4 × 9, bend the index finger (the fourth finger along) on your left hand. This leaves three fingers to the left of it, and six fingers to the right (I'm counting thumbs as fingers here).

Everything to the left of the missing finger counts as 10, everything to the right a single digit.

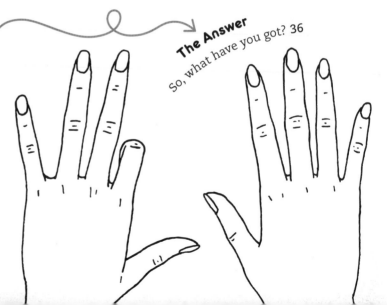

The Answer
So, what have you got? 36

Multiplying by 11

Let's Work It Out!

Multiplying by 11 often hurts the brain; it should be simple because it is just one more than ten. Well this little trick will help.

Whatever two-digit number you want to multiply by 11, add the figures together, and put them in the middle of themselves.

If they add up to anything over nine, add the first digit onto the first number and stick the remainder in the middle.

So 11 x 45 is: 4 (4+5) 5
= 495

So 11 x 29 is: 2 (2+9) 9 = 2 (11) 9
= 319

Did You Know?

When Englishman Thomas Austin moved to the state of Victoria in south-east Australia he found he couldn't hunt rabbits – the creatures were unknown in the country. So in 1859 he introduced just 12 pairs of rabbits into the local habitat. Nature (and multiplication) took its course, and soon there were so many rabbits in Victoria that two million could be killed without halting the growth of the species. The rabbits devastated native crops and altered the ecosystem of the entire continent.

The Binomial Man

Multiplying terms in algebra can look like an intimidating task, but by using the simple 'F O I L' mnemonic device you can complete the problem ... and also create a Binomial Man who smiles back at you!

Let's Work It Out!

In algebra, a binomial is an expression that consists of two terms – 'bi' meaning 'two', and 'nomial' meaning 'term' – separated by a plus or minus sign. Multiplying two binomial expressions can be similar to the multiplication of numbers. 'F-O-I-L' (First, Outside, Inside, Last) is an acronym to remember a set of rules that will help you perform this multiplication. To FOIL, as it were, you must multiply together each of the following as shown below:

As you combine the variables, draw lines to connect them, as shown in the illustration of the robot on the right – you see, he really is smiling.

When you have drawn the Binomial Man, you know you have completed all your multiplications. Now all you need to do is combine the like terms and you are done!

F = First ⟶ O = Outside ⟶ I = Inside ⟶ L = Last

The Maths

So, lets take the equation $(x + 3)(x - 2)$ and multiply the binomials.

1. Multiply the first (F) term of each binomial together: $(x \times x = x^2)$
2. Multiply the outer (O) terms together: $(3 \times -2 = -6)$
3. Multiply the inside (I) terms together: $(x \times -2 = -2x)$
4. Multiply the last (L) term of each expression together: $(+3 \times x = 3x)$

The Binomial man smiles!

BINOMIAL MAN
IS MADE OF
F O I L!

The Answer

List the four results of 'F O I L' in order.

First: $x \times x = x^2$

Outside: $x \times -2 = -2x$

Inside: $3 \times x = 3x$

Last: $3 \times -2 = -6$

Then combine the like terms to give you:

$$x^2 - 2x + 3x - 6$$

$$x^2 + x - 6$$

Multiplying Multiples

You've mastered multiplying numbers 1 to 11, what happens when the numbers keep getting larger? Well, there's a trick for that too. With this handy formula you can start practising multiplying two digit numbers, and soon you'll be able to do it all in your head.

Let's Work It Out!

The magic formula for $ab \times cd$ is:
$(a \times c), ((a \times d) + (b \times c)), (b \times d)$.

Give the letters 'a', 'b', 'c' and 'd' to the numbers in the order they appear.
Now, how do I solve 12×23?

Step 1 $a \times c$
a (1) $\times$ c (2), which gives us:
$1 \times 2 = 2$

Step 2 $(a \times d) + (b \times c)$
a (1) $\times$ d (3) + b (2) $\times$ c (2), which gives us:
$3 + 4 = 7$

Step 3 $b \times d$
b (2) $\times$ d (3), which gives us:
$2 \times 3 = 6$

The Maths

The formula uses the same idea as a table, but just loses the zeros. Again the numbers have their place value based on the order they are written.

x	10	2	
20	200	40	240
3	30	6	+ 36
			276

Rather than writing 200, the 2 is just written in the hundreds space. Our quick method follows all the right steps but just simplifies it so you can do it in your head.

The Answer

Putting the numbers back into the formula gives us 2, 7, 6, or in other words (or numbers): 276. If any number becomes ten or bigger, starting from the right, carry the number back into the next column. For 18×19:

$a \times c$	$(a \times d) + (b \times c)$	$b \times d$
1	$9 + 8 = 17$	$8 \times 9 = 72$
$1 + 2$ (carried over from middle column) $= 3$	$17 + 7$ (carried over from right-hand column) $= 24$	2

⬅ ⬅ ⬅ ⬅

Which gives: 342

Did You Know?

Zerah Colburn was born in 1804, the son of a farmer from Vermont, USA. By the age of eight, he was giving mathematical exhibitions in England where he was asked by a member of the audience to compute 8 to the sixteenth power (8^{16}). He gave the correct answer 281,474,976,710,656 in about 30 seconds, and astounded the audience. Sadly, Zerah's incredible calculating abilities waned as he aged.

The Trigonometry Tree

The maths of triangles, also known as trigonometry, can help you work out the height of a tree without having to climb up it with a tape measure! Once you know how to do it you can apply this nifty bit of maths, not just to trees, but to everything you see.

Let's Work It Out!

Standing at the bottom of the tree, walk away, counting your paces.

When you get 25 paces away, let's call this 25m (82ft), turn round and sit on the ground. Point your arm at the top of the tree and take a guess at the angle that your arm is at.

Let's say it is 50° – if directly up is 90° and along the floor is 0°.

By doing this you have created a right-angled triangle and because you know the length of one of the sides – and one of the angles – you can work out the height.

As we know the angle and we know the length of the Adjacent side, and we want to calculate the length of the Opposite side, the equation we need is:

$$\text{Tangent } (50°) = \text{Opposite/Adjacent} = \text{Opposite/25m (82ft)}$$

Most Smartphones these days have a scientific calculator. If you have one, this will give you the answer to the Tangent of 50°. Type in 50 and hit the 'TAN' button. This gives 1.19. Let's add this into the equation.

$$1.19 = \text{Opp}/25\text{m (82ft)}$$
$$\text{Opp} = 1.19 \times 25\text{m (82ft)}$$

The Answer → The height of the tree is 29.75m (97.58ft).

The Maths

To calculate different triangle measurements there are three useful equations:

SOH: Sine = $\dfrac{\text{Opposite}}{\text{Hypotenuse}}$

CAH: Cosine = $\dfrac{\text{Adjacent}}{\text{Hypotenuse}}$

TOA: Tangent = $\dfrac{\text{Opposite}}{\text{Adjacent}}$

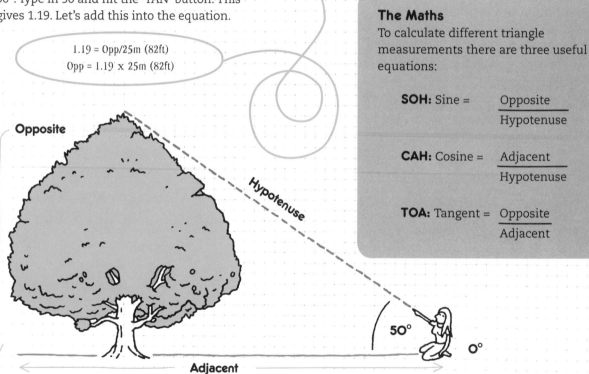

Opposite

Hypotenuse

50°

0°

Adjacent

Baby, It's Cold Outside – or Is It?

Going on a family holiday to an exotic destination? But are those temperatures in Celsius or Fahrenheit? Deciding what to pack is tricky enough already as the two units do not coincide (except when it's minus 40!). But relax, that bag will get packed, here's how to do a quick conversion.

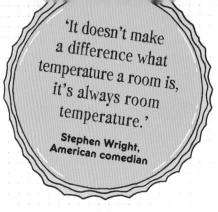

Let's Work It Out!

So, how do I convert a temperature in Celsius to Fahrenheit? For example, what is 24°C in Fahrenheit?

Step 1
Multiply the °C by 1.8:
$$24 \times 1.8 = 43.2$$
OR
Divide by 5 and multiply by 9:
$$24 \div 5 = 4.8$$
$$4.8 \times 9 = 43.2$$

Step 2 Add 32 to the answer from Step 1.

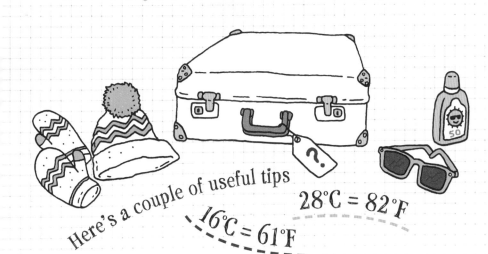

Here's a couple of useful tips

28°C = 82°F

16°C = 61°F

The Maths

The Fahrenheit scale was proposed in 1724 by German physicist Daniel Gabriel Fahrenheit (who also invented the mercury thermometer), and is based on a zero value representing the freezing point of brine. Between 1743 and 1954, the Celsius (or Centigrade) scale used the freezing and boiling point of water as its basis. Although scientists have since altered this definition, it has remained the temperature scale of the metric system, and coincides at intervals with the Kelvin scale, the measure for temperature in the International System of Units. As they were unrelated on their formations, there is little correlation between Celsius and Fahrenheit – apart from the fact they are equal at -40° – hence the need for a handy conversion method.

The Answer ⟶ 43.2° + 32 = 75.2°F

To do this in reverse (convert °F to °C), take away 32, and then divide by 1.8. (If you are just looking for a quick approximation, you can always use 2 instead of 1.8.) Or, once you have subtracted 32, divide by 9 and multiply by 5.

Who Turned the Lights Out?

Each planet in our solar system orbits the Sun due to gravity, and in turn most planets have moons in orbit around them. But with all of these huge celestial bodies moving across the sky, how can it be that the Earth's tiny Moon is able to cover the Sun and cause a solar eclipse? Here's how ...

Let's Work It Out!

A solar eclipse occurs when the Moon comes between the Earth and Sun while revolving, and blocks the Sun partially or fully. It occurs only at new moon and stops the light of the Sun from reaching the Earth. In fact, the Moon and the Sun are in conjunction as seen from Earth, so the Sun is completely invisible. The Moon does not have its own light because it relies on the Sun's light reflected by Earth, so the skies become dark during an eclipse. The time that the Sun is completely covered by the Moon is called totality.

A total solar eclipse is not noticeable until the Sun is more than 90% covered by the Moon, while at 99% coverage, daytime lighting resembles local twilight.

Did You Know?

THE SUN IS 400 TIMES LARGER THAN THE MOON!

The longest duration for a
total solar eclipse is 7½ minutes.

Total solar eclipses happen
about once every 1½ years.

Local temperatures often drop 3°C
(5°F) or more near totality.

150,000,000km

375,000km

The Maths
The diameter of the Sun is approximately 1,391,000km (864,327 miles),
and the diameter of the Moon is 3,475km (2,159 miles). If we divide the
diameter of the Sun by the diameter of the Moon, the answer is roughly
400. So, in effect, the Sun is about 400 times larger than the Moon.

Now let's look at distance. The Earth is approximately 150,000,000km
(93,205,679 miles) from the Sun, while the Moon is about 375,000km
(233,014 miles) away from the Earth. If we divide the distances in the
same way, we find that the Sun is 400 times further away from the
Earth as the Moon.

The Answer
As a result, the
Sun and Moon
both appear to
be similar in
size, and the
Moon can block
out the Sun.

Mystery Angles

A triangle is the only shape you can make using three straight lines. Basic facts about this three-sided wonder have been around roughly since 300BC. Maybe that is why maths teachers around the globe expect us to be able to calculate unknown angles, as we have had over 2,300 years of practice!

Let's Work It Out!

Can you find the missing angle?

Step 1
Add together the angles that you know.
    $30° + 80° = 110°$

Step 2
Subtract the sum from Step 1 from 180°.
    $180° - 110° = 70°$

30°

80°

The interior angles of a triangle will always add up to 180°

The Maths

The interior angles of a triangle will always add up to 180°. Try it for yourself! Cut out a paper triangle and tear off all three corners. Put the points together and line up the edges, and you will see the paper edge will make a straight line – and the angles in a straight line add up to 180°.

There are three names given to triangles that tell us how many sides or angles are equal: there can be three, two or no equal sides/angles.

The Answer
There you have it – the missing angle is 70°.

Isosceles triangles have two equal sides and two equal angles.

Scalene triangles have no equal sides and no equal angles.

Equilateral triangles have three equal sides and three equal angles (always 60°).

Did You Know?

Perhaps the most famous triangle is the Bermuda or Devil's Triangle, an area of the North Atlantic Ocean made by joining points in Bermuda, Puerto Rico and Miami, Florida. Since the early 20th century the area has seen numerous mysterious disappearances of aircraft and seagoing vessels, such as the Douglas DC-3 aircraft and its 32 passengers and crew that went missing in 1948, or USS CYCLOPS and its crew of 309 that vanished after leaving Barbados in 1918. Could it be paranormal activity? Aliens? Atlantis? Or maybe just plain old bad weather and human error are to blame. Your guess is as good as mine.

23

Happy Birthday Probability

How likely is it for you and a friend to share a birthday? The answer is more likely than you might expect. In fact, what has become known as 'the birthday problem' shows that the chance of sharing a birthday with someone in a group as small as the players on a football pitch, is definitely more likely than not.

Let's Work It Out!

In the real world events cannot be predicted with complete certainty. The best we can do is say how likely they are to happen using the idea of probability.

The coin toss is an obvious example. When a coin is tossed there are two possible outcomes: heads or tails. The probability of the coin landing on either heads or tails is 1 in 2.

Now let's think bigger: what is the probability that two people playing in a football match (two teams of eleven plus the referee) share the same birthday?

Let's imagine the referee trots out on to the pitch. He's on his own. Then the captain of the home team comes out. What is the probability that these two players do not have the same birthday?

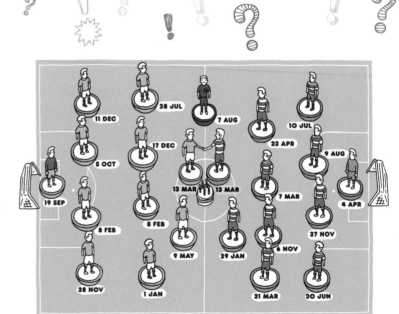

The Maths

Regardless of what day the referee's birthday is, the captain's could be on any of the remaining 364 days. So the probability of them not matching is: $^{364}/_{365}$ or, in percentage terms, a 99.72% probability that they do not share the same birthday.

When the goalkeeper comes out his birthday could fall on any of the other 363 days, so to get the probability so far we need to multiply these together to give us a $(^{364}/_{365} \times {}^{363}/_{365}) \times {}^{100}/_{1} = 99.17\%$ probability that none of these three players have the same birthday.

When everyone else comes out, if we calculate in the same way as before and keep going until all 23 players are on the pitch, the last probability will be $^{343}/_{365}$.

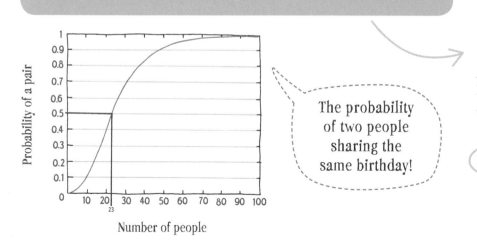

Probability of a pair

Number of people

The probability of two people sharing the same birthday!

The Answer

To allow all these maths-loving football fans to watch the game instead of tapping away at their calculators, the probability that no one on the pitch shares the same birthday is:

$$(^{364}/_{365} \times {}^{363}/_{365} \times {}^{362}/_{365} \times \ldots {}^{343}/_{365}) \times {}^{100}/_{1} = 49.27\%$$

And due to the laws of probability, the probability that two players do share the same birthday is

$$100 - 49.27 = 50.73\%.$$

What a result!

How to Tip!

All around the world, tipping for good service in restaurants is the norm ... and it is something a lot of people don't know how to work out. So, let's work it out! Using this handy tip you will now be able to work out that extra 15% without getting in a muddle.

FIND 15% OF THE FINAL AMOUNT

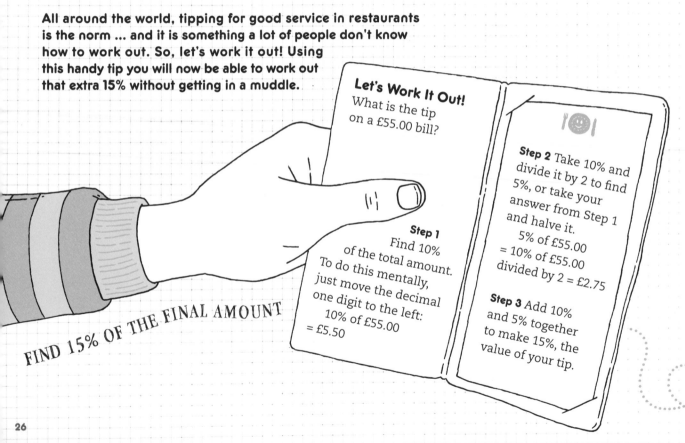

Let's Work It Out!
What is the tip on a £55.00 bill?

Step 1 Find 10% of the total amount. To do this mentally, just move the decimal one digit to the left:
10% of £55.00 = £5.50

Step 2 Take 10% and divide it by 2 to find 5%, or take your answer from Step 1 and halve it.
5% of £55.00 = 10% of £55.00 divided by 2 = £2.75

Step 3 Add 10% and 5% together to make 15%, the value of your tip.

15% **10%** **5%**

HOW'S THAT FOR A HOT TIP?

The Answer

£5.50 + £2.75 = £8.25

The Maths

When using percentages, your total amount is always equal to 100%. To find 10%, you need to divide your value for 100% by 10, which can be done by shifting the decimal. You can find 1% by dividing your original total by 100, or shifting the decimal two spots to the left. Once you have 10% and 1%, you can calculate any percent of the total, like 27% or 62% or 78% …

Thunderbolt and Lightning, Very, Very Frightening

We have all been there, home alone when a storm approaches. Bright lightning fills the sky, and then we hear it – a huge crack of thunder. Then we start to wonder, how close was that? Is the storm coming nearer or moving further away? Luckily maths is at hand and can answer our question and take our minds off the storm at the same time.

Let's Work It Out!

How can you tell how far away a lightning strike was?

Step 1 Straight after seeing a flash of lightning, time (in seconds) how long it takes before you hear the thunder.

Step 2 Divide the time by three to calculate the distance away in kilometers (five to calculate the distance away in miles).

Okay, so the lightning flashes and you start to count in seconds, you get to ten before the thunder arrives.

The Answer
$10 \div 3$ for km ($\div 5$ for miles) = 3km (2 miles)

1 – 2 – 3 – 4 – 5

To give seconds their correct length you need to allow an extra beat between each number, some people say 'Mississippi' in between, while others say 'thousand' or 'elephant'.

'Like a sudden flash of lightning, the riddle was solved. I am unable to say what was the conducting thread that connected what I previously knew with what made my success possible.'

German mathematician, Carl Friedrich Gauss (1777–1855)

Did You Know?

Roy Cleveland Sullivan was a US park ranger in Shenandoah National Park in Virginia. Between 1942 and 1977, Sullivan was hit by lightning on seven different occasions and survived all of them. For this reason, he gained the nickname 'Human Lightning Conductor' or 'Human Lightning Rod'.

The Maths

The lightning you see is moving at the speed of light. Light is very fast, covering 299,792km (186,282 miles) per second. That will make the thunder seem to travel at a snail's pace, as it only travels at 1,236km (768 miles) per hour. You will see the lightning as it strikes, but thunder on the other hand has a slight delay. If we take the 1,236km (768 miles) per hour and divide by 60, we can calculate that sound moves 21km (13 miles) per minute. Dividing by 60 again, we get 0.32 km (0.2 miles) per second, and this tells us that every 3 (5) seconds is about a kilometer (mile). 1,236km (768 miles) per hour × $\frac{1}{60}$ × $\frac{1}{60}$ = 0.32km (0.2 miles) per second.

Super Speedy Recipe Converter

A friend's mum has made you a delicious meal and you ask for the dessert recipe so you can make it yourself. The problem? Her recipe is for 12 and there are only going to be eight at your family dinner. You have two choices: 1. eat the dessert every day until it is gone, or 2. adjust the recipe. You decide on option 2, after all, who doesn't like a bit of fraction work?

Let's Work It Out!

Step 1 Find the amount by which you want to decrease (or increase) the recipe expressed as a fraction.

To do this, make your desired amount the top number (the numerator), and the original amount the bottom number (the denominator).

$$8/12$$

Reduce these numbers to their lowest terms. Both are divisible by four, which gives:

$$8 \div 4 = 2$$
$$12 \div 4 = 3$$

So in this case you want to make ⅔ of the original recipe.

Step 2 Multiply all the volumes and amounts in the recipe by your fraction; this can be done by multiplying by the numerator, and then dividing by the denominator.

The Answer

So if the recipe says 200g flour: $200g \times ⅔ = 200 \times ⅔ = ^{400}/_3 = 133.333g$. Then complete for all ingredients. Watch the cooking times though, as there is not an exact calculation for that!

The Maths

Fractions split a 'whole' into parts, with the denominator representing the total number of parts.

For example, I can order one pizza, but when it arrives it is split into six parts. Together, these parts make one pizza. The numerator stands for the number of parts you want.

Here you want to split the recipe into three parts (divide by three), and you want two of those parts (multiply by two).

COOKING IS GOOD FOR THE BRAIN

Heads or Tails?

Place your bets, ladies and gentlemen, place your bets. When a coin is flipped, which side is it going to land on? With this simple trick, it's easy to work out ...

Let's Work It Out!
How often will a flipped coin land on heads?

Step 1 Decide on the likelihood of the desired outcome.

Number of heads on a coin = 1.

Step 2 Decide how many different outcomes there are.

There are two sides of a coin = 2 possible outcomes.

Step 3 Desired outcomes/total outcomes = the probability of the event occurring.

DON'T FLIP OUT

WORK IT OUT

The Maths

To find the probability of an event occurring, or not occurring, we simply use the formula:

$$\text{Probability} = \frac{\text{number of desired outcomes}}{\text{total number of outcomes}}$$

The answer can be written as a fraction, a decimal or as a percentage.

The 'number of desired outcomes' is asking how many positive results there are. For example, if I wanted to throw an even number on a die, I would be happy with a two, a four or a six, so there are three desired outcomes.

The 'total number of outcomes' means how many possible results are there. Once again on a normal die, I could get a one, two, three, four, five or six, so there are six possible outcomes.

So the probability of throwing an even number of a die would be:

$$^3/_6 = {}^1/_2 = 0.5 = 50\%$$

The Answer

A coin will land on heads 50% of the time, or one in two flips. Remember this is an average, so in a small sample it may appear that a coin will land on heads or tails more frequently than that, but in general, if you flipped a coin 10,000 times, it would land on each side about 5,000 times.

You're so Mean!

Working out the mean, or average, is a very useful statistical tool that can be used every day. By working out the average, we can make sure everybody gets the same fair deal and no one is short-changed.

Don't get mean – get EVEN!

$$11 + 12 + 14 + 16 + 17 = 14 + 14 + 14 + 14 + 14$$

Let's Work It Out!

What is the mean of the following numbers: 11, 12, 14, 16 and 17?

Step 1 Add up all of the numbers:

$$11 + 12 + 14 + 16 + 17 = 70$$

Step 2 Divide the sum by how many numbers you added up:

$$11, 12, 14, 16, 17 = 5$$

The Maths

The mean, or average, represents a flattening or even distribution of all the numbers in a given sample. You spread the numbers out so that each group has the exact same value.

In this case I had five numbers, giving me five groups. I want each group to have the same share of 70, so I have to spread 70 out evenly between them.

This way all five groups contain the same number.

The Answer
$$70 \div 5 = 14$$

Vital Statistics

Statistics can often tell us which team is most likely to win the football match on Saturday or who is going to win an election. But how are these statistics figured out? It can be a complicated process, but with a good understanding of the basic principles and some common sense, it is very likely you'll be able to work out some statistics of your own.

Let's Work It Out!

In order to interpret simple statistics, we need to know a few key terms:

1. mean
2. median
3. mode
4. range
5. standard deviation.

Let's now give ourselves a group of numbers: 1, 5, 5, 6, 8. Even this small group can provide us with some statistical analysis, but how do we work it out?

1. The **mean**, or average, represents a flattening or even distribution of all the numbers in a given sample.

$$(1 + 5 + 5 + 6 + 8) \div 5 = 5$$

2. In a set of numbers arranged from lowest to highest, the median is the number exactly in the middle. In our group of five numbers, the middle number or **median** is 5.

3. **Mode** refers to the number that occurs the most often in a set of data. For that same set of data, we see that the mode is equal to 5.

4. The **range** is the difference between the highest and lowest value for a data set: the highest number is 8, the lowest number is 1, and so our range is equal to 7.

5. The **standard deviation** of a sample tells us how variable our answers were. If the standard deviation is small, it tells us that all of the numbers were close to the mean.

Hold on to your hats, this one needs a few steps:

Step 1 Calculate the mean.
We can see from above this is equal to 5.

Step 2 Figure out the difference between each number and the mean:
$$(1 - 5) = -4; (5 - 5) = 0;$$
$$(5 - 5) = 0; (6 - 5) = 1;$$
$$(8 - 5) = 3$$

Step 3 Calculate the square of each of the answers from the last step:
$$-4^2 = 16; 0^2 = 0;$$
$$0^2 = 0; \ 1^2 = 1; 3^2 = 9$$

Step 4 Work out the sum of the squares from Step 3:
$$16 + 0 + 0 + 1 + 9 = 26$$

Step 5 Divide the answer by the sample size (five numbers) minus one:
$$26 \div (5 - 1) = 26 \div 4 = 6.5$$

Step 6 Calculate the square root of the answer:

The Answer

$$\sqrt{6.5} = 2.55$$

So far ... so what? In a normal distribution, 68% of the data should be within one standard deviation in either direction of the mean, and 95% will fall within two standard deviations, and this tool is useful in deciding how things should be distributed – examiners will use this to determine grade boundaries. It is also used in population analysis, sports, and as a measure of risk in stock market fluctuations.

I Want to be Alone ...

To the uninitiated algebra can look like hieroglyphics. But like interpreting hieroglyphs, if you dust off the letters in your equation and move them around so you can get them on their own, then you can work out what they mean.

Let's Work It Out!

How can I isolate x in x - 3 = 5?

To get x by itself, we need to move the '-3' from one side of the equation to the other. When it moves it must become the opposite operation, and what is the opposite of subtracting three? Adding three.

$$x - 3 = 5$$
$$x = 5 + 3$$

The Maths

In algebra, when you want to move a letter or number from one side of the = sign to another, you have to use the opposite operation. Adding and subtracting are opposites – if I add a number and then take the same number away they have cancelled one another out. The same is true for multiplying and dividing.
If I divide a number by itself, it will always equal 1. Dividing or multiplying a number by 1 will always equal the same number. When you know that whatever you do to one side of an equation you have to do to the other, it is easy to solve this sort of problem.

So what about this one? How can I isolate x in $2x + 3 = 15$?

Step 1 Subtract 3 from both sides of the equation:

$2x + 3 - 3 = 15 - 3$

$2x = 12$

Step 2 Divide both sides by 2:

$2x \div 2 = 12 \div 2$

$x = 6$

Now try these:

$x - 4 = 6$

$x + 1 = 9$

$3x = 18$

$4x - 2 = 14$

$7x + 10 = 59$

For the answers see page 112.

Make sure you mind your 'X's and 'Y's

The history of algebra began in ancient Egypt and Babylon, where people learned to solve linear ($ax = b$) and quadratic ($ax^2 + bx = c$) equations, as well as indeterminate equations such as $x^2 + y^2 = z^2$, whereby several unknowns are involved. The ancient Babylonians solved arbitrary quadratic equations by essentially the same procedures taught today. They also could solve some indeterminate equations.

Did You Know?

The Answer

$x = 8$

How Fast Is He Running?

Millions of us watched Usain Bolt win the gold medal for the 100m at the London 2012 Olympic Games. Some of us know his winning time was 9.63 seconds. But how fast was he moving? Calculating speed can be a fun conversation topic, from debating the skill of football players to watching a snail in the garden. Luckily there is a simple formula that can help you find a speed in any situation.

Let's Work It Out!
Speed = distance/time

Step 1 Count in seconds/minutes how long it takes for your runner to cover a set distance.
Step 2 Estimate how many metres/ kilometers (feet/miles) the distance was.
Step 3 Divide distance by time to find metres (feet) per second/minute/hour.

So at the London Olympics Usain covered 100m (328ft) in 9.63 seconds.

The Maths
Miles per hour can be tricky to estimate, as most distances we actually see will be less than a mile. But, if we know the number of feet, we can convert feet to miles. Remember that the time it takes to run a mile will be a lot of seconds, so we also need to change our time. Using fractions to represent the same value (for example 1 mile = 5,280ft), and the idea that a unit divided by itself is essentially 'cancelled out', we can take our feet per second and change to mph. So 3ft per second expressed as miles per hour is equal to: $\frac{3}{5,280} \times 3,600$ or (just over) 2 miles per hour.

The Answer

100m (328ft) ÷ 9.63 sec = 10.38m (34ft)/second

To convert to kilometers per hour, divide by 1,000 to change metres into kilometers, and then multiply by the number of seconds in an hour (60 (seconds in a minute) × 60 (minutes in an hour)).

$$\frac{10.38}{1,000} = 0.01038 \times 3,600 = 37.37\text{km per hour}$$

Usain set his World Record for the 100m in 2009 at the Athletics World Championships in Berlin, Germany, when he ran the race in 9.58 seconds. Can you work out his speed?

Did You Know?

- A sneeze can exceed a speed of 161km (100 miles) per hour.
- A cough can reach a speed of 97km (60 miles) per hour.
- Domestic pigs can average a top speed of 17.7km (11 miles) per hour.

For the answer see page 112.

Formula, Formulae

Whether it is one or many, when it comes to rearranging a formula, we tend to make mistakes. We looked at speed on pages 40–41, but what if we now wanted to find the time it would take to travel 100km (62 miles) when travelling at 80km (50 miles) per hour? How can we take a simple formula, and rearrange it to isolate any one of the variables?

Let's Work It Out!

Let us look at our formula for speed:
speed = distance ÷ time ($s = \frac{d}{t}$).

How can I rearrange the formula for time?

Step 1 Rewrite the formula in the shape of a triangle.
Step 2 Cover the 't' (or whichever variable you want to find).
Step 3 Read out the formula left behind.

You can see that the formula for t is:

$$t = \frac{d}{s}$$

By moving your finger over the d, you can see the formula for distance is:

$$d = s \times t$$

When the letters left are on the same line, you multiply. In this example our distance is 100km (62 miles) and our speed is 80km (50 miles) per hour.

$$t = 100 \, (62) \div 80 \, (50)$$

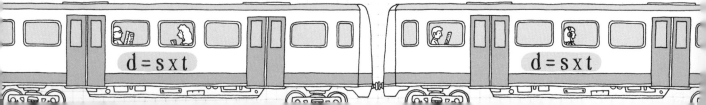

d = s x t   d = s x t

The Maths

To find the value of t from $s = \frac{d}{t}$, first you have to get the t out of the denominator position. You can do this by multiplying both sides by t, which gives us:

$$st = \frac{dt}{t}$$

We know that $\frac{t}{t} = 1$, and $d \times 1 = d$, so st = d. To get t by itself, we need to move the s. The opposite of multiplying is dividing, so we will divide both sides by s, giving us:

$$\frac{st}{s} = \frac{d}{s}$$

Knowing that $\frac{s}{s} = 1$ and $t \times 1 = t$, leaves us with $t = \frac{d}{s}$. We can go through all of these steps each time, or use the triangle method as a time-saver.

TIME IS OF THE ESSENCE!

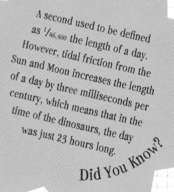

A second used to be defined as $\frac{1}{86,400}$ the length of a day. However, tidal friction from the Sun and Moon increases the length of a day by three milliseconds per century, which means that in the time of the dinosaurs, the day was just 23 hours long.

Did You Know?

The Answer → → → So it will take us 1¼ hours to travel 100km (62 miles) at 80km (50 miles) per hour.

d = s x t

d = s x t

Amazing Area

You want to buy enough paint to decorate your bedroom, but don't want a cupboard full of unfinished tins when the job is done. How do you know how much to buy? Knowing how to work out the area of a room, floor or wall, is a useful skill to have – so pay attention!

Helpful Hint
If you want to get technical, you can now find the area of all the windows and doors in the room and subtract them from your total to get the exact amount of paint required, but we might save that lesson for a later date!

Let's Work It Out!
What is the total area of wall in my dining room?

Step 1 Measure the length (or base) of the wall, from corner to corner.

Step 2 Measure the width (or height) of the same wall.

Step 3 Multiply length by width.

Step 4 Repeat for all the walls in the room.

Step 5 Add the area of all the walls together.

HOW DO YOU WORK OUT THE AREA?

$(4 \times 2) \times 2 + (5.5) \times 2$
$= (8 \times 2) + (11 \times 2)$
$= 16 + 22 = 38 m^2$

Area is calculated in square units!

Did You Know?

You can also use this method to find the area of compound shapes. Break the shape down into rectangles and square to find the areas of the different sections.

For example, the shape on the right easily breaks down into a rectangle and a square.

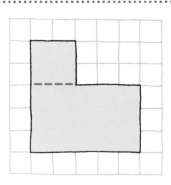

Area of the rectangle =
5cm × 3cm = 15cm²
Area of the square =
2cm × 2cm = 4cm²

Total area =
15cm² + 4cm² = 19cm²

The Maths

Area is calculated in square units, such as square metres and feet. To find how many complete square metres you have to visualize how many 1m × 1m squares you could fit on the wall; multiplication is an easy way to do this for us.

A rectangle that is 2m × 1m (6½ft × 3ft) has two 1m × 1 m squares or 2m² (6½ ft × 3ft = 19.5 one-foot squares or 19½ sq. ft). If the rectangle is 3m (10ft) × 2m (6½ft), there would be six 1m × 1m squares, or 6m² (10 × 6½ = 65 one-foot squares or 65 sq. ft). Knowing that 2m × 1m = 2m², and 3m × 2m = 6m², we can see that all you have to do is multiply length by width to find the area.

The Answer

Two of the walls of dining room measure 4m × 2m and the other two measure 5.5m × 2m. To calculate total area:

(4 x 2) x 2 + (5.5 x 2) x 2 =
(8 x 2) + (11 x 2) = 16 + 22 =
38m²

Va Va Volume

Now we've worked in two dimensions (area, see pages 44-45) you are ready for all three. Volume is not only used every time we measure out a liquid, it also helps us figure out how much space a wardrobe will take up in the corner. Volume takes us to the third dimension; who wants to be a square when you could be cubed?

Let's Work It Out!

How do we work out the volume of this box?

Step 1 First measure the length, width and depth of the object.

Step 2 Multiply all three numbers together.

Step 3 Don't forget to add units: working in metres gives us cubic metres (m³); feet will be cubic feet (cu. ft).

3-D IS SO MUCH COOLER THAN 2-D

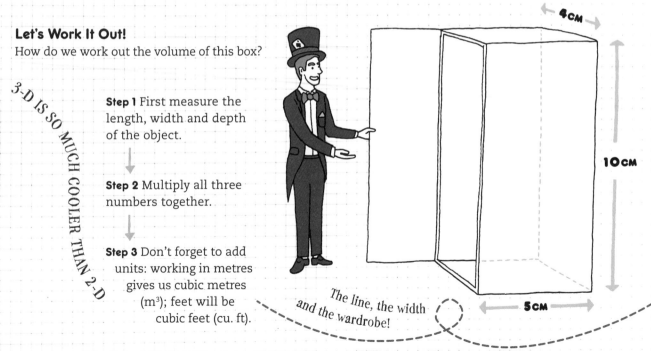

The line, the width and the wardrobe!

4CM

10CM

5CM

46

The Maths

In area we try to work out how many squares there are in an object; in volume we are trying to find the number of cubes.

Imagine you had an empty box and wanted to know how many identical cubes would fit inside. You could sit there and count, but if it is a big box, that could take a while! Volume refers to how much space a 3-D object takes up, and the formula of length × width × depth will answer just that.

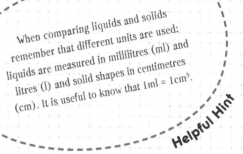

When comparing liquids and solids remember that different units are used: liquids are measured in millilitres (ml) and litres (l) and solid shapes in centimetres (cm). It is useful to know that $1ml = 1cm^3$.

Helpful Hint

The Answer

In this example:
$$5cm \times 4cm \times 10cm$$
$$= 20cm^2 \times 10cm$$
$$= 200cm^3$$

Did You Know?

Archimedes (c.287–c.212BC) is regarded by many as the greatest mathematician of antiquity. One famous story tells of how Archimedes was engaged by King Hiero II to determine whether a crown he had commissioned had been made entirely of the gold he had supplied, and not been substituted for cheaper metal. Archimedes then made a discovery while in the bath: he noticed that the more he sank into the water, the more the water rose. All he had to do was submerge the crown and measure how much water was displaced to determine its volume. The density of the crown could be obtained by dividing the mass of the crown by the volume of water displaced. Archimedes was so excited he jumped out of the bath and ran naked through the streets shouting 'Eureka!'

Going Round in Circles

These days it seems we spend a great deal of time running round in circles, but just how much energy are we wasting? How far is the distance round a circle? Circles are lines that have been bent around until the ends join up. Working out the properties of circles always makes me hungry because many of the calculations involve the use of pi (π).

Let's Work It Out!

How do I find the distance around a circle? In maths, this distance has a name – the circumference.

The life of pi = 3.14159265359 ...

Step 1 Find the diameter of the circle. The diameter is the distance straight across and passing through the centre of the circle. It can also be described as two times the radius, with the radius being the distance from the centre to the edge of the circle.

Step 2 Multiply by π; if you do not have a calculator you can use 3.14.

So let's try and work out the circumference of a circle with a diameter of 38cm (15in).

The Maths

As circles do not have any straight lines, they require a constant, π or pi. If you were to try to find the area, you would find that you could not fit squares perfectly into the shape, and even estimating becomes difficult as a circle's rounded edges make it impossible to tell the exact boundaries. It all goes back, once again, to Archimedes (see page 47). He found an upper and lower boundary for π, and by the 17th century 35 decimal places had been calculated. Currently, the number of digits for π continues to grow; we cannot be expected to memorize them all, but most calculators have a button that stores enough numbers for us.

Circumference = 38cm (15in) × π = 119cm (47in)

This can also be expressed as 2πr where r = radius.

Have you ever noticed that when athletes line up to run the 400m they are always staggered at the start? The closer the lanes are to the edge of the track the further the athletes in those lanes have to run, so mathematicians need to determine the exact distance of each lane. By calculating the circumference of the semi-circles and adding these to the straight section of the track, they can adjust the starting lines to make the race fair.

Did You Know?

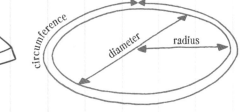

circumference

diameter   radius

Pythagor-who?

How long a ladder do I need to reach to my roof? How much shorter would my journey be if I cut through the park rather than walking around it? Every day we encounter problems that involve right-angled triangles, and thankfully we have Pythagorean Theorem or Pythagoras' Theory to help us answer these questions.

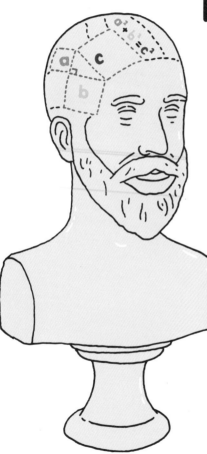

Let's Work It Out!

Pythagorean Theorem is based on the idea that if you have a right-angled triangle and you made a square on each of the three sides, the biggest square would have the same area as the other two squares put together. This can be expressed as:

$$a^2 + b^2 = c^2$$

Where c is the longest side of the triangle and a and b are the other two sides.

Knowing this to be true, if we know the lengths of two sides of a right-angled triangle, we can find the length of the missing one. The longest side of a right-angled triangle is called the **hypotenuse**.

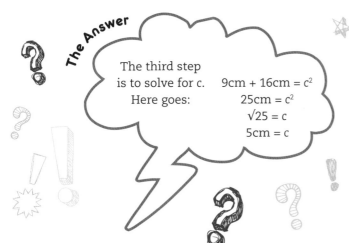

The third step is to solve for c. Here goes:

$$9cm + 16cm = c^2$$
$$25cm = c^2$$
$$\sqrt{25} = c$$
$$5cm = c$$

So what is the length of the hypotenuse of a right-angled triangle with sides that measure 3cm and 4cm?

Step 1 Write out the formula:
$a^2 + b^2 = c^2$

Step 2 Substitute the values for the lengths of the known sides into the formula for a and b:

$$3cm^2 + 4cm^2 = c^2$$

The Maths

The formal definition of the theorem is as follows: 'In a right-angled triangle the square of the hypotenuse is equal to the sum of the squares of the other two sides.' Referring to the diagram inside Pythagoras' head (see diagram far left), we find that:

The area of square a + area of square b = area of square c

Once we find the area of c, we can convert this into the length by calculating its square root; as most of your answers will not be whole numbers, the ($\sqrt{}$) on your calculator will do the work for you.

Shop 'Til You Drop

You are watching your favourite TV show and the adverts come on. You are just about to change channels but it's sale time and the discount on that new TV sounds amazing! But how much does it actually cost? Is it a good deal, or is it still cheaper at the other store down the road? With a little bit of practice, you can become a savvy shopper in no time!

Happy shopping!

Calculate the discount

Let's Work It Out!

How do I calculate 20% off a TV that costs £500?

Step 1 Find 10% of the total price by moving the decimal point one place to the left.

10% of £500.00 = £50.00

Step 2 Double it, as 10% + 10% = 20%

£50.00 + £50.00 = £100.00

Step 3 Subtract the discount from the original price to get the sale price.

The Maths

Calculating a sale price is based around the same skill we used to calculate a tip (see pages 26–27). To do that, we worked out 5% by finding 10% and then dividing that amount by two, we then added the amount we wanted to tip on to the bill.

Here we are using the same operation: we treat the original price as 100%, and can find 10% and 1% by moving the decimal point one or two places to the left. The difference when calculating a discount is that you subtract the value from the total rather than add it on. So go on, hit the shops – there are bargains to be had and maths to be done.

The Answer

£500 - £100 =
£400

What Day Is It?

Without having a calendar in front of you, it may seem an impossible task to decide on the best day for your party. This handy trick helps you work out on which day of the week a date in the future will fall: everyone loves a party on a Friday or Saturday night, but if it turns out to be a Monday, you might just end up celebrating on your own.

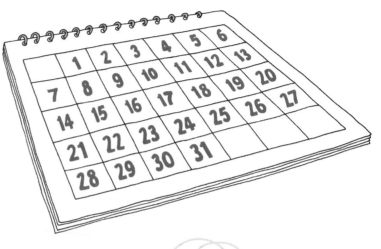

Did You Know?

Before the Gregorian calendar, most countries relied on the Julian calendar, which was introduced by Julius Caesar in 45BC. It was in common use until the 1500s but created an error of one day every 128 years. The Gregorian calendar was proposed by Aloysius Lilius, a physician from Naples, Italy, and was adopted by Pope Gregory XIII in accordance with the instructions from the Council of Trent (1545–63) to correct for errors in the older Julian calendar. It was decreed by Pope Gregory XIII in a papal bull on 24 February 1582. The reformed calendar was adopted later that year by a handful of countries, with other countries adopting it over the following centuries. It is now the most widely accepted civil calendar.

Let's Work It Out!

What day of the week will it be 47 days from Wednesday?

Step 1 Divide the number of days in the future by seven and list how many days are left over:

$$47 \div 7 = 6, \text{ remainder } 5$$

Step 2 Count the number of the remainder days forwards from Wednesday:
  1 day ahead is Thursday
  2 days ahead is Friday
  3 days ahead is Saturday
  4 days ahead is Sunday
  5 days ahead is Monday.

Predict the future

The Answer

This means the day of the week 47 days ahead of Wednesday will be a Monday – not the best day for a party, but at least you can impress your friends with this amazing trick!

Amazing Trick

The Maths

The basis of this mystery uses the fact that seven days from today will be the same day of the week from which you started: seven days ahead from a Tuesday will be another Tuesday. If we take the total number of days and divide it by seven, the whole number represents how many weeks there are until we end up on the exact same day in the future. The remainder can then be used to count the relevant number of days forward.

You can use this to find past dates as well, but remember the remainder will be days before, so go backwards instead of forwards when working out the day of the week.

More practically, you can use this trick to predict what day of the week an actual event will be on. If you want to know what day of the week your next birthday is going to be on, count how many days away it is and use the same method. Make sure you remember that old saying: 'Thirty days hath September, April, June and November. February has twenty-eight alone, all the rest have thirty-one.' And don't forget about leap years!

Kaprekar's Constant

Dattaraya Ramchandra Kaprekar (1905–86) was an Indian mathematician who discovered several number constants (or cycles) during his life. No one paid a great deal of attention to his work as he was a schoolteacher rather than a scholar, but today these numbers provide us with some interesting and fun calculations.

Let's Work It Out!

Step 1 Pick any four-digit number that is made from at least two different numbers (so not 1,111 or 2,222).

Step 2 Arrange the numbers in increasing order.

Step 3 Arrange the numbers in decreasing order.

Step 4 Subtract the number from Step 2 from the number from Step 3.

Step 5 Use the number you obtain and repeat the steps above.

For example:

1. Choose number
   3,141
2. Increasing
   1,134
3. Decreasing
   4,311
4. Step 3 - Step 2
   4,311 - 1,134 =
   3,177

2. Increasing
   1,377
3. Decreasing
   7,731
4. Step 3 - Step 2
   7,731 - 1,377 =
   6,354

2. Increasing
   3,456
3. Decreasing
   6,543
4. Step 3 - Step 2
   6,543 - 3,456 =
   3,087

2. Increasing
   0378
3. Decreasing
   8,730
4. Step 3 - Step 2
   0378 - 8,730 =
   8,352

2. Increasing
   2,358
3. Decreasing
   8,532
4. Step 3 - Step 2
   8,532 - 2,358 =
   6,174

Once you have reached 6,174 and you go through the steps again you get: 7,641 - 1,467 = 6,174 And this number keeps repeating – hence the name 'constant'.

Mr Kaprekar also has a type of number named after him: a Kaprekar number for a given base is a non-negative integer, the representation of whose square in that base can be split into two parts that add up to the original number again. For instance, 45 is a Kaprekar number, because $45^2 = 2025$ and $20+25 = 45$.

The Maths
Each number in the sequence uniquely determines the next number in the sequence. Since there are only a finite number of possibilities, eventually the sequence must return to a number it hit before, leading to a cycle. So any starting number will give a sequence that eventually becomes a cycle. This also works for three-digit numbers, but this time the constant will be 495. Give it a try!

These numbers lead to a cycle

Taxonomy Fun

Taxonomy is the classification of organisms. Hold on a minute, does this sound like fun? Well not for everyone perhaps, but this trick should make you smile.

Let's Work It Out!

Step 1 Pick a number between 1 and 10.

Step 2 Multiply by 9 (For a quick way to do this see page 10.)

Step 3 Add the digits.

Step 4 Subtract 5.

Step 5 Match the number with the corresponding letter of the alphabet (1 = a, 2 = b, and so on).

Step 6 Think of a country that starts with that letter.

Step 7 Think of an animal that starts with the last letter of the country.

Step 8 Think of a colour that starts with the last letter of the animal's name.

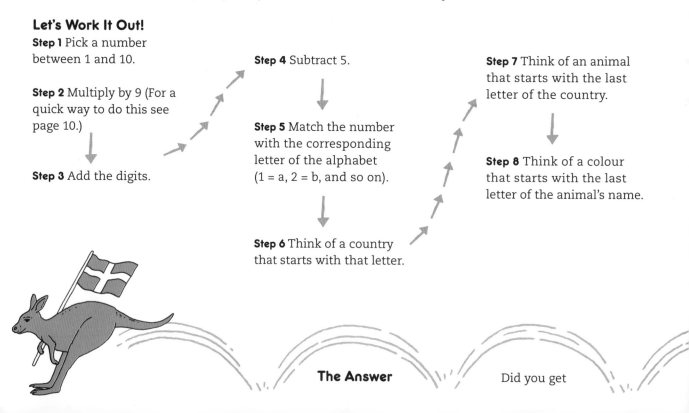

The Answer   Did you get

GIVE IT A TRY!

The Maths

The trick here is that the number-letter association is always 'd', no matter what number you choose. This is because any single-digit number multiplied by 9 will give an answer with the digits adding up to 9, which will lead to the answer 4, and the fourth letter of the alphabet is 'd'. Denmark is the only country in Europe that begins with the letter 'd' (the only other global options are: Djibouti, Dominica, Dominican Republic).

For those who choose Denmark, the majority of people will think of the animal 'kangaroo' for the letter 'k' because of the easy colour association of the last letter with 'orange'.

an   orange kangaroo   from Denmark?   Most people will!

Palindrome Numbers

Palindromes are words or phrases that read the same forwards as backwards, for example, 'race car', or 'A man, a plan, a canal, Panama'. You can write the letters from left to right, or right to left, and you get exactly the same thing! And you can do this with numbers too!

1.

Let's Work It Out!
Step 1 Pick a number.

Step 2 Reverse it.

Step 3 Add the two numbers together.

Step 4 If this is not a palindrome, repeat Steps 2 and 3.

2.

So let's try this:
1. 723
2. 327
3. 723 + 327 = 1050

3.

This is clearly not a palindrome, so repeat Steps 2 and 3.

The Maths

In fact, about 80% of all numbers under 10,000 solve in four steps or less. About 90% solve in seven steps or less. A rare case, number 89, takes 24 steps to become a palindrome.

In fact it's been found that all numbers less than 10,000 will produce a palindrome in this way with one bizarre exception – the number 196. Although it's been taken through hundreds of thousands of reverse-and-add steps, leading to giant 80,000-digit numbers, no palindrome has yet been found. Numbers like this are called Lychrels.

Here are some more palindrome numbers:
1. 87
2. 78
3. 87 + 78 = 165
4. 165 + 561 = 726
5. 726 + 627 = 1,353
6. 1,353 + 3,531 = 4,884

1. 132
2. 231
3. 132 + 231 = 363

1. 2,346
2. 6,432
3. 2,346 + 6,432 = 8,778

> **READS THE SAME BACKWARDS AS WELL AS FORWARDS.**

> What other palindromes do you know?

Did You Know?

Palindromes date back at least to 79AD, as one was found at Herculaneum, one of the cities buried by ash following the eruption of Mt Vesuvius in that year. This palindrome, written in Latin, is known as the Sator Square and reads: 'Sator Arepo Tenet Opera Rotas' ('The sower Arepo holds works wheels'). It is remarkable for the fact that the first letters of each word form the first word, the second letters form the second word, and so forth. Hence, it can be arranged into a word square that reads in four different ways: horizontally or vertically from either top left to bottom right or bottom right to top left.

Divisibility Rules Okay!

Division may not be your cup of tea, but don't panic as there are some simple rules that will allow you to test whether one number is divisible by another, without having to do too much maths. Now doesn't that sound better?

Let's Work It Out!

DIVIDE AND CONQUER!

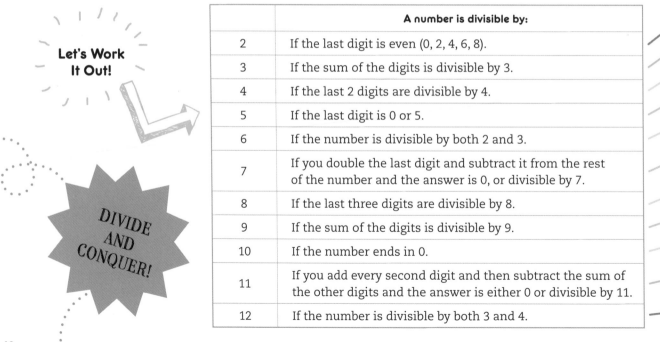

	A number is divisible by:
2	If the last digit is even (0, 2, 4, 6, 8).
3	If the sum of the digits is divisible by 3.
4	If the last 2 digits are divisible by 4.
5	If the last digit is 0 or 5.
6	If the number is divisible by both 2 and 3.
7	If you double the last digit and subtract it from the rest of the number and the answer is 0, or divisible by 7.
8	If the last three digits are divisible by 8.
9	If the sum of the digits is divisible by 9.
10	If the number ends in 0.
11	If you add every second digit and then subtract the sum of the other digits and the answer is either 0 or divisible by 11.
12	If the number is divisible by both 3 and 4.

Number	Example	The Maths	The Answer
2	128 129		Yes No
3	381 217	(3 + 8 + 1 = 12; 12 ÷ 3 = 4) (2 + 1 + 7 = 10; 10 ÷ 3 = 3.33)	Yes No
4	1,312 7,019	(12 ÷ 4 = 3) (19 ÷ 4 = 4.75)	Yes No
5	175 809	.	Yes No
6	114 308	(Even; 1 + 1 + 4 = 6; 6 ÷ 3 = 2) (Even; 3 + 0 + 8 = 11; 11 ÷ 3 = 3.66)	Yes No
7	672 905	(2 × 2 = 4; 67 - 4 = 63; 63 ÷ 7 = 9) (2 × 5 = 10; 90 - 10 = 80; 80 ÷ 7 = 11.43)	Yes No
8	109,816 216,302	(816 ÷ 8 = 102) (302 ÷ 8 = 37.75)	Yes No
9	1,629 2,103	(1 + 6 + 2 + 9 = 18; 1 + 8 = 9) (2 + 0 + 1 + 3 = 6)	Yes No
10	220 221		Yes No
11	1,364 25,176	((3 + 4) - (1+6) = 0) ((5 + 7) - (2 + 1 + 6) = 3)	Yes No
12	648 524	3: 6 + 4 + 8 = 18; 18 ÷ 3 = 6 Yes 4: 48 ÷ 4 = 12 Yes 3: 5 + 2 + 4 = 11; 11 ÷ 3 = 3.66 No 4: 4 ÷ 4 = 1 Yes	Yes No

Hair has the highest rate of mitosis (cell division). An average hair grows 0.3mm a day and 1cm (¼in) per month.

Did You Know?

Number Tricks 1

Number tricks are a great way to amaze your friends and family. Give these ones a try.

Trick 1

1. Think of a number less than 10.

2. Double it.

3. Add 6.

4. Halve it.

5. Subtract the original number from the answer.

Trick 2

Think of a number.

Subtract 1.

Multiply by 3.

Add 12.

Divide by 3.

Add 5.

Subtract the original number from the answer.

Is your number 3?

Trick 4
1. Pick two single-digit numbers.

2. Pick one of them and double it.

3. Add 5. + 5

4. Multiply by 5. x 5

5. Add the second number.

6. Subtract 4. - 4

7. Subtract 21. - 21

Trick 3

Think of a number. ▷ Multiply it by 3. ▷ Add 45.

Subtract the original number from the answer. ◁ Divide by 6. ◁ Double it.

The Answer

Is your answer the two numbers you started with?

Is your number 8?

The Answer

Is your number 15?

The Answer

Number Tricks 2

Are your friends and family amazed by your maths skills? Here are some more tricks to keep you going.

The Answer

Is the answer the month and day of your birth?

Trick 5

Write down your house number.

Multiply by 2.

Add the number of days in a week.

Multiply by 50.

Add your age.

Subtract 365 (number of days in a year).

Add 15.

Trick 6

Multiply the number of the month of your birthday by 5.

Add 7.

Multiply by 4.

Add 13.

Multiply by 5.

Add the day of your birthday.

Subtract 205.

Is the answer 9?

Trick 7

1. Enter into a calculator any number that consists solely of the number nine repeated.

2. Multiply it by any number.

3. Write down the number on paper.

4. Add together the individual digits in the answer.

5. Add the answer digits together.

(If not …

Trick 8

1. Choose a number from 1 to 10.

2. Double it.

3. Add 2 to the result.

4. Divide that number by 2.

5. Subtract the original number from the answer in Step 4.

Is the answer 1?

Is your answer your house number and then your age?

… keep adding the new answer digits together and eventually they will add up to 9.)

Mind-boggling Numbers

When we first start to count the numbers are small and easy to keep track of – we use our fingers to help us. But as we get older we find that numbers get bigger and bigger and bigger. And once we become interested in astronomy some numbers are too big to comprehend. Luckily scientific notation, also known as standard form, exists in maths to help us get large numbers down to manageable proportions.

Let's Work It Out!

How can I write a number in scientific notation? For example 4,560?

Step 1 Add a decimal point after the first digit: So 4,560 becomes 4.560

Step 2 Count how many times you moved the decimal over. This will be the exponent in your final answer.
$$4.560 = 3$$

Step 3 Take your number and multiply it by 10 to the exponent from Step 2; as long as you don't break any significant figures rules (see page 76), you usually do not need to include all the zeros.

The Answer
4.560×10^3 or 4.56×10^3

The Maths

Scientific notation is a way to write really big (or small) numbers by using an exponent of the number 10.

$10^1 = 10$
$10^2 = 10 \times 10 = 100$
$10^3 = 10 \times 10 \times 10 = 1000$, and so on.

This becomes useful when you want to try to condense a large number down. Let's say your answer was 3,450,000,000. That would be the same as saying 3.45 × 1,000,000,000. And 1,000,000,000 is the same as saying 10^9, as 10 × 10 × 10 × 10 × 10 × 10 × 10 × 10 × 10 = 1,000,000,000. So our answer can be written as 3.45×10^9, or 3.45 billion.

To extract a number from scientific notation, you need to move the decimal. So if the exponent is 6, move the decimal 6 places to the right, adding zeros if you run out of numbers. If the exponent is negative, you move the decimal to the left, making the number smaller.

Still boggled?

Avogadro's Number, 6.022×10^{23}, is the number or molecules in one mole of a substance. For those of you who may not be a lover of chemistry, that means there are 6.022×10^{23} individual molecules of H_2O in every 18g (⅓ oz) of water. Hope you're thirsty!

Did You Know?

It's Hip to be Square

You know what shape a square is, but what is squaring in maths? When you are asked to square a number, you multiply the number by itself. If you had a square, both the length and the width would be the same number – see how they are related?

English schoolmaster and mathematician Charles Hutton published the first table of squares up to 25,400 in the year 1781, commissioned by the Board of Longitude. After learning of the table, mathematicians and amateurs gleaned many facts from it, such as the simple observation that squares always end in the digits 0, 1, 4, 5, 6, or 9, never 2, 3, 7 or 8.

Did You Know?

Let's Work It Out!
What is 15 × 15?

Step 1 Take the first digit, and multiply it by its next highest digit:
$1 \times (1 + 1) = 1 \times (2) = 2$

Step 2 Multiply the two fives together:
$5 \times 5 = 25$

MULTIPLY THE NUMBER BY ITSELF!

The Maths

The formula we are using here is N × (N + 1). Using the same idea as when we were multiplying two-digit numbers, we know that we need to multiply our tens and ones. So for the ones, when we multiply 5 by 5 we will always get 25; if we multiply 5 by 10 we get 50, and there two of them, which added together will make 100; and when we multiply 10 by 10 we get 100 more – so we get a total of 225.

But what if we want to find the square of a number that doesn't end in 5, like 19?

Multiply the number by itself!

Step 1 Find the difference between the number you want to square and the nearest multiple of 10; for 19 the nearest boundary is up one digit to 20.

Step 2 Depending on the direction of the boundary, now count the relevant number of the digits in the other direction; for 19 we would now count one digit down to 18.

Step 3 Multiply the numbers from Steps 1 and 2 together: 20 × 18 = 360. (To do this without a calculator multiply by 10 and then double the answer.)

Step 4 Add the square of the difference from Step 1:
1 × 1 = 1
360 + 1 = 361

	1	2	3	4	5	6	7	8	9	10
1	1	2	3	4	5	6	7	8	9	10
2	2	4	6	8	10	12	14	16	18	20
3	3	6	9	12	15	18	21	24	27	30
4	4	8	12	16	20	24	28	32	36	40
5	5	10	15	20	25	30	35	40	45	50
6	6	12	18	24	30	36	42	48	54	60
7	7	14	21	28	35	42	49	56	63	70
8	8	16	24	32	40	48	56	64	72	80
9	9	18	27	36	45	54	63	72	81	90
10	10	20	30	40	50	60	70	80	90	100

By manipulating our numbers so that we always have a multiple of ten, we simplify these problems so we can do them in our heads. Remembering too that we can break down any larger number into two or more numbers that multiply to make it as well, the question gets easier and easier.

Summation of Sums

Have you ever stared at your pile of change at the checkout and wondered whether you have enough money to buy the milk (that you definitely need) and the chocolate bar (that you don't)? Can you afford the ice cream and the crisps (you really don't want to have to choose)? If you learn this useful trick then you really won't need to use a calculator to work out the answer.

SUM LIKE IT HOT!

Let's Work It Out!

How can I solve 81 + 78?

Step 1 We want to use tens as much as possible, so round the second number to the nearest 10.
78 + 2 = 80

Step 2 Add this number to the first number:
80 + 81 = 161

Step 3 To get your answer, add or subtract the difference between the second number and the number you rounded it to.

The Pascaline was possibly the first mechanical adding device actually used for a practical purpose. It was built in 1643 by Blaise Pascal to help his father, Etienne, a tax collector, with the tedious activity of adding and subtracting large sequences of numbers. However, it was not until 1820 that French inventor and entrepreneur Charles Xavier Thomas designed, patented and manufactured the first commercially successful mechanical calculator, called the Arithmometer.

Did You Know?

The Answer
This gives you:

How about this one:
62 + 53?
62 + 50 (- 3) = 112
112 + 3 = 115

161 - 2 = 159

The Maths
Just like the multiplication trick on pages 80–81, we can rewrite any number as the combination of numbers that add or subtract to make it. So the number 7 can be written as 3 + 4, or 10 - 3, as both operations are equal to 7. When we look at two-digit numbers, we may find it easier to rewrite them as an addition or subtraction from the nearest ten, as most of us can quickly add ten onto a number. We can then finish the problem by adding or subtracting the difference. Give it a go next time you are at the corner shop.

Or this: 97 + 35?
97 + 30 (- 5) = 127
127 + 5 = 132

A HANDY LITTLE TRICK

Don't be so Negative

Now that we've mastered adding up, the next basic skill we need is taking away. However, double negatives can be confusing when you are completing a maths problem. But don't worry, this little number trick will get you feeling positive about negatives ...

Let's Work It Out!

How do I subtract a negative number, for example 6 - (-4)?

When you have two negatives, or a subtraction and a negative together, this is the same as a positive, so subtracting minus 4 is the same as adding 4:

$$6 + 4$$

The Answer
When you add the numbers together you get:
6 + 4 = 10

BANK

OH NO, I'VE GOT A NEGATIVE BALANCE!

Positive Numbers

-10 -9 -8 -7 -6 -5 -4 -3 -2 -1 0 1 2 3 4 5 6 7 8 9 10

Negative Numbers

The Maths

In the order of operation (see pages 78–79) subtraction always comes last, so the numbers must stay where they are. Using a number line is a nice way to keep track visually.

Addition moves you towards the right, but every time you see a subtraction or a negative sign you need to change direction. When you have a basic subtraction, you turn once, so you move to the left, or down the number line, but when you have two negatives, you turn and then turn again, so back up the number line you go!

Now you know that subtracting a negative number is the same as adding, remember that adding a negative number is the same as subtracting.

$$\text{So } 7 + (-3) = 4$$

Did You Know?

Rhesus is a protein that occurs on red blood cells. People who have the protein are known as Rhesus positive, and those without it are called Rhesus negative. The highest incidence of Rhesus negative is among the people of the Basque region that lies between France and Spain, where it is around 30%; for the rest of Europe this falls to around 16%. By contrast, among Asian and African people it affects less than 1% of the population.

Rules for Multiplying

POSITIVE × POSITIVE = POSITIVE

POSITIVE × NEGATIVE = NEGATIVE

NEGATIVE × POSITIVE = NEGATIVE

NEGATIVE × NEGATIVE = POSITIVE

Significant Figures

We all know that builders' estimates can add up – usually to extra money in their piggy banks. But when we are working something out, often we only need a rough idea. To do this we can round numbers up or down, or choose the number of decimal places we want to use. Another way to produce an estimate is to use significant figures.

Let's Work It Out!

What is the number 368,249 to three significant figures?

Step 1 For 368,249, the '3' is the most significant digit because it tells us that the number is three hundred thousand and something, but as we want three significant figures in our answer, we need to move along to the '8'.

Step 2 Now we need to look at the number that follows the '8'. As this is a '2' the rounding rules tell us that we should round down rather than up.

The rules for rounding up are:
- If the next number is 5 or more, we round up.
- If the next number is 4 or less, we do not round up.

The Maths

In maths we round off numbers to a certain number of significant figures; the most common are one, two and three.

The rules for significant figures are:

1. All non-zero numbers (1, 2, 3, 4, 5, 6, 7, 8, 9) are always significant.

2. All zeros between non-zero numbers are always significant, eg. 30,245.

3. All zeros that are simultaneously to the right of the decimal point and at the end of the number are always significant, eg. 501.040.

4. All zeros to the left of a written decimal point and are in a number greater than or equal to ten are always significant, eg. 900.06.

You can use significant figures for decimals too. For 0.0000058763, the '5' is the most significant digit, because it tells us that the number is five millionths and something. The '8' is the next most significant, and so on.

Learn the rules!

The Answer

So 368,249 to three significant figures would be 368,000.

77

The Order of Things

You've probably already noticed, but maths is a peculiar animal – you can do a calculation in different ways and come up with a new answer each time. Fortunately, as you might expect in maths, there is a right way to do things, and this is called the 'Order of Operations'.

Let's Work It Out!

How do I solve 4 × (3 + 4) ÷ 14 + 5?

Step 1 Complete the work in the brackets:
4 × (7) ÷ 14 + 5

Step 2 Complete all multiplication and division from left to right:
4 × 7 = 28
28 ÷ 14 + 5
28 ÷ 14 = 2
2 + 5

Step 3 The final stage is to complete all addition and subtraction from left to right.

PLEASE EXCUSE MY DEAR AUNT SALLY!

PEM
DAS

The Answer
$2 + 5 = 7$

The Maths

When solving a multi-step problem, always complete using the convention represented by the following acronym: 'PEMDAS'.

This stands for:

Parentheses (Brackets): if any portion of the question is enclosed, focus your efforts here first.

Exponents (Indices/Orders): these are the little numbers written in superscript immediately after a number. They tell you how many times to multiply a number by itself; also known as a 'power'.

Multiplication and Division: these are tied for importance, so complete any and all from left to right.

Addition and Subtraction: these should also be completed from left to right.

'Do not worry about your difficulties in mathematics, I assure you that mine are greater.'

Albert Einstein, German-born physicist

Helpful Hints for Multiplying

With a bit of practice, you can multiply numbers in less time than it takes to turn your phone to calculator mode. Not only will this impress your friends, you will also give your brain a workout. Practice these helpful hints to beat anyone in a multiplication showdown.

Let's Work It Out!

x 4 the number four can be broken down into 2 × 2, so if we need to multiply by four we can actually multiply by two, and then by two again. In other words, double the number, and then double it again.

x 5 Five is half of ten. If you can multiply by ten and halve it, it is the same as multiplying by five.

x 6 Multiply by three and then double it.

x 12 Double the number and add it to ten times the original.

x 14 Multiply by seven and then double it.

x 16 Multiply by eight and then double it.

x 18 Multiply by 20 (or by ten and double it), then subtract the number twice. Alternatively, times by nine and double it, now that you are an expert at your nine times tables!

The Answer
See page 112.

Test Yourself
Try to answer these questions in your head:
1. $4 \times 9 =$
2. $11 \times 8 =$
3. $6 \times 4 =$
4. $7 \times 12 =$
5. $5 \times 14 =$
6. $16 \times 9 =$
7. $3 \times 18 =$
8. $11 \times 12 =$
9. $8 \times 6 =$
10. $18 \times 9 =$

Practice makes perfect!

2 3 7 9

1 4 0 6

8 5

The Maths

When multiplying, the order of multiplication does not affect your final answer. This allows you to break a large number down into the smaller numbers that multiply to make it. I could break 16 into 4 × 4, which I can then split into 2 × 2 × 2 × 2. Instead of multiplying by 16, I can just double the number four times. Or multiply by four and then double it twice. Or, as we have just seen, times by eight and then double it – the possibilities are endless!

Breeding Like Rabbits

If you were lucky enough to be wealthy in Italy during the Middle Ages you had quite a lot of time on your hands, and in the absence of television, the Internet and other modern pastimes, people tended to do quite a lot of thinking. This was certainly the case for Leonardo Pisano Bigollo (c.1170–c.1250), also known as Fibonacci, the son of a rich merchant, who spent his time thinking about numbers.

Let's Work It Out!

Fibonacci travelled extensively and studied the Hindi–Arabic numeral system and in 1200 published a book titled *Liber Abaci*. One of the problems considered in this book involved the growth of a population of rabbits. The question he asked was:

Suppose you go to an uninhabited island with a pair of newborn rabbits (one male and one female) who mature at the age of one month, have two offspring (one male and one female) each month after that, and live forever. Each pair of rabbits matures in one month and then produces a pair of newborns at the beginning of every following month. How many pairs of rabbits will there be in a year? The solution begins as follows:

1. At the end of the first month, they mate, but there is still only one pair.

2. At the end of the second month the female produces a new pair, so now there are two pairs of rabbits in the field.

3. At the end of the third month, the original female produces a second pair, making three pairs in all in the field. This process then continued according to Fibonacci's sequence …

The Maths

Month	1	2	3	4	5	6	7	8	9	10	11	12
Pairs of rabbits	1	2	3	5	8	13	21	34	55	89	144	?

The sequence generated by the rabbit problem is called the Fibonacci Sequence and has many applications in both mathematics and in nature. Expressed as a formula the rule is:

$$X_n = X_{n-1} + X_{n-2} \cdots$$

1

2

3

4

5

Where:
x_n is term number 'n'
x_{n-1} is the previous term (n - 1)
x_{n-2} is the term before that (n - 2) ...

HOW MANY RABBITS ARE THERE?

In the Prime of Life

Weren't the Ancient Greeks great? They did a lot of the hard work in maths so we don't have to. Eratosthenes of Cyrene (c.276BC–c.195BC) was a Greek scholar who invented a neat way to work out prime numbers from 1 to 100.

The Maths

What Eratosthenes did was to create a simple algorithm to find the prime numbers up to a given limit. It does this by starting from each prime number and identifying all composites of that prime. As this uses a sequence of a numbers with the same difference equal to that prime this is more efficient than using trial and error to find each prime number.

'Sift the Two's and Sift the Three's, The Sieve of Eratosthenes. When the multiples sublime, The numbers that remain are Prime.'

Anonymous

Eratosthenes was also a poet, an astronomer and a geographer. He was the first person to use the word 'geography' in Greek and invented the discipline of geography as we understand it.

Did You Know?

Let's Work It Out!

Prime numbers are those numbers (greater than 1) that cannot be divided by any number except themselves and one.

1. Write out the numbers from 1–100 in ten rows of ten.

2. Cross off number one, because all primes are greater than one.

3. Number two is a prime, so we can keep it, but we need to cross off the multiples of two (i.e. even numbers).

4. Number three is also a prime, so again we keep it and cross off the multiples of three.

5. The next number left is five (because four has been crossed off), so we keep it and cross off multiples of this number.

6. The final number left in the first row is number seven, so cross off its multiples.

7. You have finished; the numbers left over (coloured in white below) on your grid are prime numbers.

The Answer

	2	3	4	5	6	7	8	9	10
11	12	13	14	15	16	17	18	19	20
21	22	23	24	25	26	27	28	29	30
31	32	33	34	35	36	37	38	39	40
41	42	43	44	45	46	47	48	49	50
51	52	53	54	55	56	57	58	59	60
61	62	63	64	65	66	67	68	69	70
71	72	73	74	75	76	77	78	79	80
81	82	83	84	85	86	87	88	89	90
91	92	93	94	95	96	97	98	99	100

Four Colour Theorem

While we all associate maths with science, what about other subjects? One of the last areas we might think of is geography, but one problem concerning the colours used on maps kept mathematicians busy for years.

Let's Work It Out!

On a political map, each neighbouring country or state needs to be a different colour so that the map is clear. In the 19th century there was also the issue of cost: the more colours used, the more expensive the map was to print. So what was the smallest number of colours needed so that neighbouring countries or states were always different colours?

A number of mathematicians took on the challenge. In the 1850s Englishman Francis Guthrie suggested four colours were sufficient, and in 1879, London barrister Alfred Bray Kempe offered a proof that was then disproved 11 years later. The problem went on to baffle mathematicians for the next 100 years.

The Maths

In maths the Four Colour theorem states that, given any separation of a plane into contiguous regions, producing a figure called a map, no more than four colours are required to colour the regions of the map so that no two adjacent regions have the same colour. Kempe's flawed 'proof' involved the idea of an 'unavoidable set' of configurations, which derived from Leonard Euler's work on geometric figures, where countries could only have up to five neighbours. He also argued that if a map were to need five colours, then removing one country would allow that map to be simplified down to four colours. He would then reinstate that country and see if he could make the configuration work again. It was this concept of reducibility that inspired Heesch, Appel and Haken.

The Answer

The answer, as it turned out, was very unmathematical indeed. In the 1960s German mathematician Heinrich Heesch began to use computers to continue Kempe's work. News of this then reached two Americans, Kenneth Appel and Wolfgang Haken of the University of Illinois. They devised a computer program that tested all possible configurations and reduced them down until they could continue no further, at which point the process was abandoned and the program began again with a different configuration. And finally in June 1976, after just under 2,000 configurations and 1,000 hours of computer time, they achieved it.

Turkey is officially, politically and geographically part of both the European and Asian continents. Its established dividing line between Asia and Europe is the Bosphorus Strait.

Did You Know?

The 'I's Have It!

The Romans' mathematical legacy is a curious one, that used letters rather than the Arabic numerals we are familiar with today. Its complexity is blamed by many for the fact that despite all their advances in other respects, no mathematical innovations took place during the Roman Empire and Republic, and there were no mathematicians of note.

A particular number has four letters. Take two letters away and you have four left. Take one more letter away and you have five left. What is the word?

Did you know?

For the answer see page 112.

Let's Work It Out!

How do you write the year 2013 in Roman numerals?

The Roman used a special method of expressing numbers based on the following symbols:

To write the year 2013 in Roman numerals you need to break it down into its constituent units, i.e. thousands, hundreds, tens and ones, and write it down in order:

2000 = MM
13 = XIII

The Answer
2013 = MMXIII

1	2	3	4	5	6	7	8	9
I	II	III	IV	V	VI	VII	VIII	IX

10	20	30	40	50	60	70	80	90
X	XX	XXX	XL	L	LX	LXX	LXXX	XC

100	200	300	400	500	600	700	800	900	1,000
C	CC	CCC	CD	D	DC	DCC	DCCC	CM	M

The Maths

The Roman system of numerals is based on symbols used by the Etruscans, a civilization based in north-west Italy from c.1200 until the beginning of the Roman republic in the 1st century BC.

Many believe that the numbers from 1–5 were based on the shape of the fingers: I represents one finger, II two fingers etc. and the oblique line of the V represents the thumb. The 'X' of the number ten represents two crossed thumbs. The symbols for the higher numbers – L, C, D and M – come from the modified symbols 'V' and 'X'.

The way the numbers are formed is based on addition and subtraction and these are the rules:

1. When a symbol appears after a larger symbol it is added:

VI = V + I = 5 + 1 = 6

2. But if the symbol appears before a larger symbol it is subtracted:

IX = X - I = 10 - 1 = 9

3. Don't use the same symbol more than three times in a row.

4. A bar placed on top of a letter or string of letters increases the numeral's value by 1,000 times.

Work out your birthday in Roman numerals!

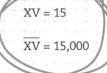

XV = 15

$\overline{XV}$ = 15,000

Squaring the Circle

Come on, step up to the oche – it's time for a game of darts. But just how big is the target you are aiming at? (Let's forget about the bullseye, you'll never hit that!) We've already worked out how to calculate the distance around a circle (see pages 48–49), so now it's time to find out how to work out its area.

Let's Work It Out!
A dartboard has a radius of 22.86cm (9in) and the bullseye in the centre has a radius of 1.27cm (½in). What is the area of the dartboard outside the bullseye.

The formula for the area of circle is:

$$\text{Area} = \pi r^2$$

The radius is the distance from the centre of the circle to the edge and is half the length of the diameter.

So to calculate the area of the larger circle we have:

$$\pi r^2 =$$
$$\pi \times (22.86\text{cm/9in})^2 =$$
$$\pi \times 522.58\text{cm (81in)} =$$
$$1,641.73\text{cm}^2 \ (254\tfrac{1}{2}\text{ sq. in})$$

The Maths

For any circle its circumference divided by its diameter is equal to 3.141592 … This relationship has been known since antiquity, but the Greek letter π (pi) was first used to describe it by Welsh mathematician William Jones (1675–1749) in 1706.

π is an irrational number and cannot be written as a fraction. Expressed as a decimal π has an infinite number of digits with no apparent pattern, although mathematicians have sought to find one by calculating π to more and more decimal places – it has now been calculated to over ten trillion (10^{13}) digits.

This gives us:

$$1{,}641.73 \text{cm}^2 \ (254\tfrac{1}{2} \text{ sq. in}) - 5.06 \text{cm}^2 \ (\tfrac{4}{5} \text{ sq. in}) = 1{,}636.67 \text{cm}^2 \ (253\tfrac{7}{10} \text{ sq. in})$$

And for the smaller circle:

$$\pi r^2 =$$
$$\pi \times 1.27 \text{cm} \ (\tfrac{1}{2}\text{in})^2 =$$
$$\pi \times 1.61 \text{cm} \ (\tfrac{1}{2}\text{in}) =$$
$$5.06 \text{cm}^2 \ (\tfrac{4}{5} \text{ sq. in})$$

To work out the area of the board outside the bullseye we need to subtract the area of the smaller circle from the larger one.

A Golden Photograph

The Golden Ratio created by Fibonacci's sequence of numbers (see pages 82–83), gives us a neat way to take better photographs by helping us with composition. So whether your camera is disposable, long lens or simply on your phone, get clicking – but make sure you remember the golden rule.

The Golden Ratio applied to a rectangle

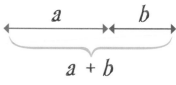

THE GOLDEN RULE

Let's Work It Out!

Ancient Greek mathematicians first studied what we now know as the Golden Ratio because of its appearance in geometry relating to pentagons and pentagrams. In 1202, Fibonacci published his sequence of numbers (see pages 82–83), and it became apparent that the further up the sequence you move, the ratio between the numbers becomes closer and closer to the Golden Ratio.

But what does this have to do with taking photographs? If you apply the idea of the Golden Ratio to a rectangle, then the most aesthetically pleasing shape is one where the ratio of the shorter to the longer sides is somewhere around 1.6 – the value of φ. And, if you divide this rectangle again by creating a square and another rectangle, the smaller rectangle will be another golden rectangle. If you carry on, this will create a spiral shape that relates to shells seen in nature that exhibit the properties of the Fibonacci sequence.

The Answer

So when you are taking a picture, imagine placing a Fibonacci spiral on top of the image. Then, position the most important element of your shot, e.g. someone's eyes, an important building, not at the exact centre of the image, but at the eye of the Fibonacci spiral, which is slightly off-centre. Try it – it really works!

Did You Know?

Almost 2,500 years ago, a Greek sculptor and architect called Phidias is thought to have used the Golden Ratio to design the statues he sculpted for the Parthenon, and the word 'phi' in his name actually inspired the naming of this number in the 20th century.

Mine Is Bigger Than Yours

Maths does not always involve absolute values. Sometimes we compare something to something else and determine their value relative to one another: Peter ran faster than John; Emma's hair is shorter than Jane's etc. These are known as inequalities. They also have their own symbols, which is good because that makes us look clever.

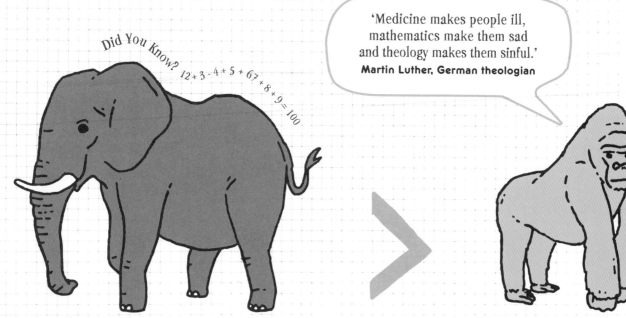

Did You Know? $12 + 3 - 4 + 5 + 67 + 8 + 9 = 100$

'Medicine makes people ill, mathematics make them sad and theology makes them sinful.'
Martin Luther, German theologian

Let's Work It Out!

When you go to the cinema, all movies are given a classification that tells you whether or not it is suitable for your age group. Let's say a movie is classified '15', which means that no one younger than 15 is allowed to watch it, how would you write that as an inequality? You may also be thinking 'That's not fair!' but that's not the sort of inequality we're talking about right now.

So to watch a movie rated '15' you have to be 15 or older. So in terms of inequalities this means you need to be:

>15

But you can also be equal to 15 ...

The Maths

We use the term inequalities to describe things that are not equal. The two most common inequalities are:
> Greater than
< Less than

Remember the narrow end always points towards the smaller number: BIG > small.

Inequalities can also include 'equals':
≥ Greater than or equal to
≤ Less than or equal to

The Answer

... which means that to watch a movie rated '15' your age must be:

≥15

Let's Factor That In

Factoring a number in maths is a bit like putting it through a mincer: you use it to find out which numbers divide into it exactly, including the number 1 and the number itself. It also helps us find out which numbers other numbers have in common.

What occurs once in every minute, twice in every moment and yet never in a thousand years? (For the answer see page 112.)

Did You Know?

Let's Work It Out!
I've got 80 lollipops but I don't know if I've got 12 or 20 children coming to the party. I don't want to have to worry about dividing them up later, so how can I divide the lollipops so the children get the highest number possible in either case?

The factors of 12 are:

1, 2, 3, 4, 6 and 12

The factors of 20 are:

1, 2, 4, 5, 10 and 20

The Maths

Here we are calculating what is called the Highest Common Factor of 12 and 20. The Highest Common Factor (HCF) of two whole numbers is the largest whole number that is a factor of both.

We can also find the HCF by multiplying all the prime factors that appear in *both* lists, which in this case is just the number two, so:

$$2 \times 2 = 4$$

Another common operation involving factors is to find the Lowest Common Multiple. The Lowest Common Multiple (LCM) of two whole numbers is the smallest whole number that is a multiple of both. You work this out by multiplying all the prime factors that appear in **either** list:

$$12 = 2 \times 2 \times 3$$
$$20 = 2 \times 2 \times 5$$
$$LCM = 2 \times 3 \times 5 = 60$$

THE
X FACTOR

The Answer

You can see here that the number four appears in both lists, and is the highest number that is common to both, so if I put four lollipops into each bag then whether the party is crowded or not, each child will get the same sugar rush.

Six of One and Half a Dozen of the Other

Fractions are all very well, but what do we do with them? I know I would like half of that cake, a third of that one, and that other one looks nice too, but I should probably only have a quarter. What on Earth does that add up to? I already know the answer is 'Too much'!

Let's Work It Out!

What is ½ + ⅓ + ¼?

There are three steps to adding fractions:

Step 1 Make sure the numbers on the bottom (the denominators) are the same.

Step 2 Add the top numbers (the numerators) over the denominators.

Step 3 Simplify the fraction if necessary.

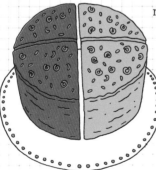

Here the denominators are different so we need to make them the same. One way to do this is to find the Lowest Common Multiple (LCM) of the three numbers (see pages 96–97). The prime factors of 2, 3 and 4 multiplied together would be:

$$LCM = 2 \times 2 \times 3 = 12$$

This means our Lowest Common Denominator needs to be expressed in twelfths, so we get:

$$^{6}/_{12} \text{ (equal to } ^{1}/_{2}) + {}^{4}/_{12} \text{ (equal to } ^{1}/_{3}) + {}^{3}/_{12} \text{ (equal to } ^{1}/_{4})$$

To find the total we need to add the numerators together.

The Maths

Subtracting fractions is done in the same way, but multiplication and division are different. To multiply fractions you do this:

1. Multiply the numerators.
2. Multiply the denominators.
3. Simplify the fraction if necessary.

And to divide them, you do this:

1. Turn the second fraction upside-down.
2. Multiply the first fraction by the second fraction.
3. Simplify the fraction if necessary.

FRACTIONS ARE FUN!

Some fractions give a recurring number like $^2/_3$, which is 0.66666 .... Others give us more interesting numbers, such as $^{617}/_{500} = 1.234$, while there are some fractions that create repeating patterns, for example, $^{152}/_{333} = 0.456456456 ...$

Did You Know?

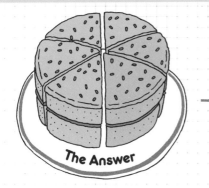

The Answer

$$^6/_{12} + ^4/_{12} + ^3/_{12} = {}^{13}/_{12} = 1^1/_{12}$$

$^{13}/_{12}$ is actually an improper fraction (the numerator is greater than the denominator), so definitely too much cake!

Euclidean Efficiency

Active in the late 4th century BC, Euclid was a Greek mathematician famous for his *Elements*, one of the most influential works in the history of mathematics. Very little is known about his life but his methods formed the basis of maths textbooks until the early 20th century. Among his numerous legacies was an algorithm for calculating the Highest Common Factor (HCF, see pages 96–97), without even using factors.

Did You Know?

Euclid's proof of his algorithm was done geometrically as algebra had not yet been invented. His algorithm is considered to be one of the best examples of an efficient algorithm.

Let's Work It Out!
What is the HCF of 36 and 15 using Euclid's Algorithm?

In Book VII of *Elements* Euclid described how to work out the HCF without listing its factors. To find the HCF of 36 and 15 carry out the following steps:

Step 1 Divide the greater number by the smaller number:

$$36 \div 15 = 2 \text{ (remainder 6)}$$

Step 2 Divide the smaller number by the remainder:

$$15 \div 6 = 2 \text{ (remainder 3)}$$

Step 3 Divide the first remainder by the second remainder:

$$6 \div 3 = 2 \text{ (remainder 0)}$$

THINK OF AN ALGORITHM AS A MATHS RECIPE!

The Answer
The last non-zero remainder is the HCF, so therefore:
HCF = 3

The Maths

An algorithm is a step-by-step process for a particular operation – a bit like a maths recipe. Another way of expressing the algorithm is like this:

For two numbers a and b:
$a \div b$ gives a remainder of r
$b \div r$ gives a remainder of s
$r \div s$ gives a remainder of t
...
$w \div x$ gives a remainder of y
$x \div y$ gives no remainder

Here y is the HCF of a and b. If the first step had produced no remainder, then b (the smaller of the two numbers) is the HCF.

Cracking the Code

Everyone loves a spy story, but being a spy is not about guns and gadgets, it is about secrets. Codes, like the substitution cipher seen here, have been used for as long as language has been written down to protect information from our enemies, famous examples being Julius Caesar, Mary Queen of Scots, and the German Enigma code during the Second World War.

Let's Work It Out!

Codes and ciphers are forms of secret communication. A code replaces entire words, phrases, or sentences with groups of letters or numbers; a cipher rearranges letters or substitutes them with other letters or symbols to disguise the message. This is called encryption or enciphering.

Here is a message you might leave for someone in your family or a friend, but what does it say?

WKH NHB LV XQGHU WKH PDW

When Julius Caesar sent coded messages to his generals his cipher used letters that were three places further along in the alphabet: so for A he used D, for E he used H, and so on. So using the key below, can you crack the code?

A	B	C	D	E	F	G	H	I	J	K	L	M	N	O	P	Q	R	S	T	U	V	W	X	Y	Z
0	1	2	3	4	5	6	7	8	9	10	11	12	13	14	15	16	17	18	19	20	21	22	23	24	25

The Answer

The message says:
'The key is under the mat'.

The Maths

To create the cipher we are adding three to our starting number, so for A this would be:

$$0 + 3 = 3$$

The number 3 represents the letter D in our cipher so we would substitute that for the letter A. E would be:

$$4 + 3 = 7$$

Here the number 7 is H, so E would become H when we were encrypting our message. When we are decoding the messages we are doing the opposite, so in this example, encryption is addition while decryption is subtraction. Using the terms of cryptology, the method used to create the code is known as the algorithm, while the table above, which is used to allow the original message (or plaintext) to be enciphered and deciphered, is known as the key.

In fact, if you look at the original encrypted message above, H is the letter that appears most often, followed closely by W. This is because E and T are the two most common letters in the alphabet, and this use of frequency analysis is one method code breakers use to crack codes.

Did You Know?

The Enigma machine was first invented in 1918 by Arthur Scherbius, a German businessman, who sold it commercially to banks. A military version of the machine was used by Nazi Germany before and during the Second World War. The Germans believed the code to be unbreakable, but code breakers first in Poland and later in the UK, helped by Alan Turing's early computer, the Bombe, managed to break the code and shorten the war by as much as two years.

Our Survey Says ...

You've done a survey and you want to do a really fancy presentation but what's the best way to present your information? Numbers by themselves are dull, but luckily there are many different types of graphs that enable us to show our findings in an exciting way.

Let's Work It Out!

Out of a group of 20 friends, these are the types of television shows they like best:

Entertainment	Food	Comedy	News	Sport
5	4	6	1	4

How can we show these as both a bar chart and a pie chart?

To make a bar chart you want to show how many people prefer each category of show. So you plot the number of people on the vertical axis and the types of show along the bottom.

Making a pie chart – a circular chart that uses sections of relative sizes to compare data – is a bit more complicated. First, for the group as a whole (20 people), you need to work out as a percentage how many people preferred each category.

Then you need to work out the angle of each section as a percentage of 360° (the number of degrees in a circle).

↓

Entertainment	Food	Comedy	News	Sport
5	4	6	1	4
5/20 = 25%	4/20 = 20%	6/20 = 30%	1/20 = 5%	4/20 = 20%
25% of 360° = 90°	20% of 360° = 72°	30% of 360° = 108°	5% of 360° = 18°	20% of 360° = 72°

Now you can go ahead and divide up your circle accordingly (you'll need a protractor).

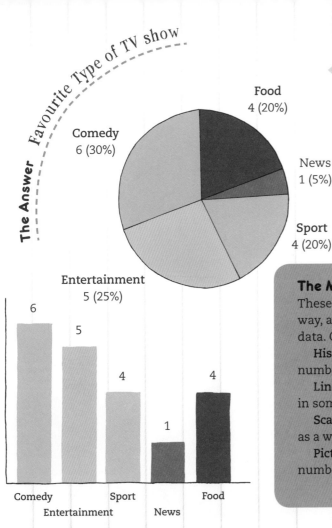

Favourite Type of TV show

Food
4 (20%)

Comedy
6 (30%)

News
1 (5%)

Sport
4 (20%)

Entertainment
5 (25%)

Comedy · Entertainment · Sport · News · Food

Tetraphobia is a fear of the number four. In China and other countries in South-east Asia, including Korea, Japan and Vietnam, the word for four sounds similar to the word for death. As a result, the number four and many other numbers containing the number four are often avoided, for example, floor numbers are skipped in buildings, and table numbers are missed out at weddings and other social occasions.

Did You Know?

The Maths

These are just two ways to show data in a graphical way, and are good for showing the relative size of data. Other types of graphs include:

Histograms: Similar to bar charts but with numbers grouped into ranges.

Line graphs: Show information that is connected in some way, e.g. change over time.

Scatter charts: Plot one set of data against another as a way to show the relationship between the two.

Pictograms: Use images to stand for a certain number of things.

Magic Squares

Have you ever tried Sudoku? These fiendish number puzzles are everywhere – in books, newspapers and online. Wouldn't you just love to work out how to do it? Well let's go back to basics and find out how to build magic number squares and you can take it from there.

Let's Work It Out!

How do you construct a square using consecutive numbers where every row, column and diagonal add up to the same number? Cover up the grid below, draw your own and see how you get on.

Step 1 Draw a 3 × 3 grid. Place a number 1 in the middle of the top row.

Step 2 Move through the other squares placing the numbers consecutively. Once you have placed a number remember to move:
* The square diagonally up and to the right when you can.
* The square below if you cannot.
* If you move off one edge of the square you must re-enter on the other side.

Step 3 When you place the 2 you are moving up and to the right, but as you are moving off the edge of the square the 2 is placed in the bottom right-hand corner. Likewise with the number 3 you are moving off the right-hand side of the square, so you must re-enter in the middle row on the left-hand side. The space for the 4 is occupied so it must be placed down below the three, and so on …

The Answer

8	1	6
3	5	7
4	9	2

As you can see each row, column and diagonal add up to 15. Can you do the same for a 5 × 5 grid? (For the answer see page 112.)

The Maths

The method used to construct this square is an algorithm, or a series of steps, known by many names including de la Loubère's algorithm, the staircase method and the Siamese method. Simon de la Loubère was a French mathematician and diplomat who brought back the method following his time spent in Siam (now Thailand) in the 17th century. He published his findings in a book, *A New Historical Relation to the Kingdom of Siam*, in 1693.

The method involves a simple arithmetic progression and works when you start with any odd number. (A normal magic square starts with the number one; a magic square can start with any positive number.) There is no algorithm for generating magic squares starting with an even number.

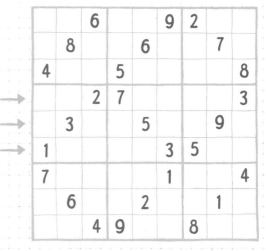

Did You Know?

Given any magic square, rotating it or reflecting it will produce another magic square. Not counting these as distinct, it is known that for 3 x 3 squares there is only one normal magic square, and for 4 x 4 squares there are 880 normal magic squares. As the size of the square increases the number of normal magic squares increases dramatically, and for 5 x 5 squares there are over 13 million normal magic squares!

DESIGN YOUR OWN SUDOKU
— GIVE IT A GO!

Mind Your 'X's and 'Y's

Simultaneous equations, or equations that involve two or more unknowns that have the same value in each equation, can be infuriating but they can come in handy. So don't get too cross, and look at those 'Y's with no fear in your eyes.

Let's Work It Out!

You've been to a café with two groups of friends and you know what the bill came to each time and who had what, but what does everything cost?

The first time you went there were six of you, six of you had the set meal, five of you had coffee, and the bill came to £37.50. The second time you went there were four people, again you all had the set meal and this time only two of you had coffee, and the bill was £23. How much does the set meal and the coffee cost? So where x denotes set meals and y denotes coffees, this gives us the equation:

$$6x + 5y = 37.5$$
$$4x + 2y = 23$$

As none of the unknowns cancel one another out (i.e. there is not a '+ x' in one equation and a '- x' in the other) the first thing to do is to give one of the unknowns the same value in each equation. So let's do that for y:

Multiply the first equation by two to give us:
12x + 10y = 75

Multiply the second equation by five to give us:
20 x + 10 y = 115

The Answer

To find a value for x, subtract the first equation from the second equation to cancel out the y:

$$(20x - 12x) + (10y - 10y) = 115 - 75$$
$$8x = 40$$
$$x = 5$$

Now substitute the value for x back into the first equation and rearrange it to give us a value for y (you can check your answer by making sure it works for the other equation too):

$$6(5) + 5y = 37.5$$
$$30 + 5y = 37.5$$
$$5y = 37.5 - 30$$
$$5y = 7.5$$
$$y = 7.5 \div 5 = 1.5$$

So the set meals cost £5 while the coffees are £1.50.

SIMULTANEOUS EQUATIONS CAN COME IN HANDY

Lunch Menu

Coffee

Set Meal

What is the total?

The Maths

Simultaneous equations require detective work: first, get one of the suspects on its own and try and get it to give up its accomplices. Use your Lowest Common Multiple (see pages 96–97) to get everyone to tell the same story, and then take one away from the other and rearrange to get the answer you want. Job done.

A Mirror Image

We see symmetry in the shapes and objects all around us. Even our earliest art creation was a perfect example – remember those butterfly pictures you made when you were young? The ones where you painted on one side of the paper and then folded it in half so the paint was transferred to the other side? I bet you didn't realize that was as much about maths as it was about art.

Symmetry is everywhere you look!

Did You Know?

Narcissus was a hunter in Greek mythology, who was renowned for his beauty and arrogance. The goddess Nemesis (you can probably guess where this is going) lured Narcissus to a pool where he saw his own reflection in the water and fell in love with it. Unable to tear himself away from his reflection he died, and the term 'narcissist' is now used to described people who are fixated with themselves.

The Answer

So what about the lines of symmetry in a circle?
A line drawn at any angle that goes through the centre is a line of symmetry, so a circle has infinite lines of symmetry.

Let's Work It Out!

So you folded your paper to make your butterfly, which makes one line of symmetry. But what about other shapes? How many lines of symmetry does a circle have?

For a shape to be symmetrical, when you fold that shape in half, the folded part must sit perfectly on top with all edges matching. But some shapes have more lines of symmetry than others.

A rectangle has two lines of symmetry

A square has four lines of symmetry:

An equilateral triangle (where all sides are the same length) has three lines of symmetry:

As you might expect, other regular polygons, also with sides of the same length have as many lines of symmetry as they have sides (try it, it's true).

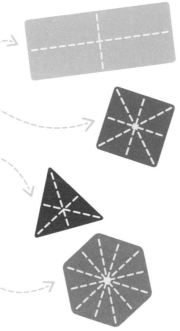

pp.38–39 I Want to be Alone

x = 10, x = 8, x = 6, x = 4, x = 7

pp.40–41 How Fast Is He Running?

37.58km (23.35 miles) per hour

pp.80–81 Helpful Hints for Multiplying

1. 36	**6.** 144
2. 88	**7.** 54
3. 24	**8.** 132
4. 84	**9.** 48
5. 70	**10.** 162

pp.88–89 The 'I's Have It!

The word is 'FIVE'.
Take away 'F' and 'E', and you get 'IV' (the Roman numeral for four). Take away 'I' and you are left with 'V' (the Roman numeral for five).

pp.96–97 Let's Factor That In

The letter 'm'.

pp.106–107 Magic Squares

17	24	1	8	15
23	5	7	14	16
4	6	13	20	22
10	12	19	21	3
11	18	25	2	9

For this grid each row, column and diagonal add up to 65.

ANSWERS!